# RAMAN SPECTROSCOPY
*Theory and Practice*

# RAMAN SPECTROSCOPY
*Theory and Practice*

Edited by Herman A. Szymanski
*Chairman, Chemistry Department*
*Canisius College*
*Buffalo, New York*

℗ PLENUM PRESS · NEW YORK · 1967

ISBN-13: 978-1-4684-3026-4    e-ISBN-13: 978-1-4684-3024-0
DOI: 10. 1007/978-1-4684-3024-0

Library of Congress Catalog Card Number 64-23241

© 1967 Plenum Press
Softcover reprint of the hardcover 1st edition    1967

A Division of Plenum Publishing Corporation
227 West 17 Street, New York, N. Y. 10011
All rights reserved

# Preface

The concept of this book — an integrated and comprehensive coverage of all aspects of Raman spectroscopy by a group of specialists — took form nearly three years ago. It made a giant stride toward realization when Dr. L. Woodward, whose outstanding work in this field had long been known to me, agreed to write an introductory chapter and made valuable suggestions concerning others who might be invited to contribute articles. However, many obstacles had to be overcome before this book finally became a reality.

It is extremely difficult to prepare a multiauthor review of the state of knowledge in a rapidly growing field in such a way that all aspects are brought up to the same date. In our case, some workers who had originally agreed to contribute articles were forced to withdraw under the pressure of new commitments, and replacements had to be found. Others were unable to complete their contributions by the deadline date, so that the publication of the book had to be rescheduled. All this tended to work to the detriment of those authors who prepared their chapters as originally scheduled. An effort was made to give the authors most affected by this an opportunity to revise their papers, but of course an arbitrary cutoff date had to be set to avoid an endless spiral of revision and updating. In view of these facts, I extend my gratitude to all who contributed to this volume, and my apologies to those who, punctual themselves, were inconvenienced by the unpunctuality of others.

How well this volume realizes the stated objective, the reader himself will best be able to judge. I hope that even a cursory insight into the editor's problems will temper his judgment.

H. A. Szymanski

*Buffalo, New York*
*December 1966*

# Contents

*Chapter 1*

# General Introduction

## L. A. Woodward

*Inorganic Chemistry Laboratory*
*University of Oxford*
*Oxford, England*

---

The object of this chapter is to provide the nonspecialist reader with a general account of the nature and theory of the Raman effect and to indicate the types of application in which it has been found to be of value. Owing to considerations of space, the treatment must necessarily be rather superficial in character, but it is hoped that this introduction may provide a broad background against which the subsequent more specialized chapters may be read. References to these will be made at suitable points.

### THE RAMAN EFFECT

Raman spectroscopy is concerned with the phenomenon of a change of frequency when light is scattered by molecules. If the frequency of the incident light is $v_0$ and that of a component of the scattered light is $v_r$, then the frequency shift $v_r - v_0 = \Delta v$ may be either positive or negative in sign. Its magnitude is referred to as a *Raman frequency*. The set of Raman frequencies of the scattering species constitutes its *Raman spectrum*.

A frequency shift $\Delta v$ is equivalent to an energy change $\Delta v/h$. It is convenient and usual to express observed results in terms of *wave numbers* instead of frequencies. A frequency $v$ is, of course, a number of vibrations per second. The corresponding quantity expressed in wave numbers (for which in this book the distinguishing symbol $\tilde{v}$ will be used) is the number of waves per centimeter, and so is related to $v$ by the equation $\tilde{v} = v/c$, where $c$ is the velocity of light. It is customary to continue to refer to the "frequency" of a spectrum line, even when the value is expressed in wave numbers. Although this is, strictly speaking, a misnomer, no serious confusion arises in practice.

1

## Historical

The naming of the Raman effect honors Sir C. V. Raman, who discovered it experimentally[1] in 1928 in the course of extended researches on molecular light-scattering. It should be noted that the effect had been foreseen theoretically by Smekal[2] in 1923. Also, at about the same time as Raman's discovery, Landsberg and Mandelstam[3] observed this effect in quartz. Raman's paper, however, represented the more thorough study, and in 1930 he was awarded a Nobel Prize.

Consider a clear substance (solid, liquid, or gas) irradiated by monochromatic light (usually in the visible region) whose frequency $v_0$ is chosen so that it does not correspond to any absorption by the sample. Almost the whole of the light will pass through the sample unaffected, but a very small part of it will be scattered by the molecules in directions different from that of the incident beam. When the scattered light is studied spectroscopically, it is found that a high proportion of it has the same frequency $v_0$ as the primary light. This constitutes the so-called Rayleigh or classical scattering, the existence of which had been known long before the Raman effect was discovered. The intensity of Rayleigh scattering is proportional to the fourth power of $v_0$. Hence, if white light is used, the blue end of the spectrum is scattered more strongly than the red. This had been recognized as the explanation of the blue color of the clear sky, which arises from the Rayleigh scattering of white sunlight by the molecules in the atmosphere.

Raman's discovery consisted of the observation that, when monochromatic light of frequency $v_0$ is used, the spectrum of the scattered light shows (in addition to the Rayleigh line) a pattern of lines of shifted frequency—the Raman spectrum. The *shifts* (Raman frequencies) are independent of the exciting frequency $v_0$ and are characteristic of the species giving rise to the scattering. The pattern on the low-frequency side of the exciting line ($\Delta v$ negative) is "mirrored" by an identical pattern on the high-frequency side ($\Delta v$ positive), but the intensities for $\Delta v$ negative are greater than those for $\Delta v$ positive, and the latter show a rapid falling off as $|\Delta v|$ increases.

## Nature of Raman Scattering

The presence of scattered lines of shifted frequencies can be understood by considering the incident light to consist of photons of energy $h\nu_0$. On collision with a molecule, a photon may be elastically scattered, i.e., without change of energy; this gives rise to the Rayleigh line. Collisions may, however, be inelastic. They may cause the molecule to undergo a quantum transition to a higher energy level, with the result that the photon loses energy and is scattered with lower frequency ($\Delta\nu$ negative). If the molecule is already in an energy level above its lowest, an encounter with a photon may cause it to undergo a transition to a lower energy, in which case the photon is scattered with increased frequency ($\Delta\nu$ positive). Thus, we see that the Raman shifts are equivalent to the energy changes involved in transitions of the scattering species and are therefore characteristic of it. Moreover, since at temperature equilibrium the population of a higher level is less than that of a lower level and falls off exponentially with the energy, we can understand why it is that the Raman shifts with positive $\Delta\nu$ are less intense than those with negative $\Delta\nu$ and why for the former the intensity falls off rapidly as $|\Delta\nu|$ increases.

In fact, it is found that Raman shifts correspond to vibrational or rotational transitions of the scattering molecule. Such frequencies, when observed by direct-absorption techniques, lie of course in the infrared spectral region. In Raman spectroscopy they can be observed (as frequency shifts) in the more convenient visible region. It must not be thought, however, that Raman spectroscopy is just an alternative method of obtaining the results that could be otherwise obtained by the infrared method, because the two differ essentially in mechanism, the Raman effect being a scattering phenomenon and the other an absorption phenomenon. In consequence, the data obtainable by the two methods are not identical. As we shall see in greater detail below, frequencies permitted in the Raman effect may be forbidden in the infrared, and *vice versa*. The two methods are thus complementary in character, and data from both should be taken into account in order to obtain the maximum amount of information in any particular case.

Raman scattering with diminution of frequency is formally reminiscent of fluorescence. The name "Stokes lines" for Raman lines with negative $\Delta\nu$ has come into general use on this account, in memory of G. G. Stokes and his pioneer work upon fluorescence. In some respects this is unfortunate, and it must not be taken to imply that the phenomena of fluorescence and Raman scattering are cognate. (Incidentally, the convention of referring to Raman lines with positive $\Delta\nu$ as "anti-Stokes" is even less felicitous!) In fluorescence, the incident photon is completely absorbed and the molecule concerned is thereby raised to an excited electronic level. After a certain lifetime in this upper state, the molecule undergoes a downward transition and thereby reradiates light of a frequency lower than that which it had absorbed. This mechanism is radically different from that of the Raman effect, in which the photon as a whole is never absorbed, but rather perturbs the molecule and induces it to undergo a vibrational or rotational transition. The essential difference is shown by the fact that fluorescence can be quenched by adding a species that is able to take away the energy of the molecule by collisions after it has absorbed the incident photon, but before it has had time to reradiate. No such quenching is possible for the Raman effect, because this does not involve the actual attainment of an excited intermediate level. We shall return to this point later in discussing the quantum-mechanical theory of Raman scattering.

### Experimental Arrangements

The overriding factor in the design of apparatus for the study of Raman spectra is the extreme feebleness of the effect. For liquid samples and light in the visible region, the total intensity of molecular scattering (including Rayleigh scattering) may be of the order of $10^{-5}$ of the incident intensity, and of this perhaps only about 1% may contribute to the Raman spectrum. For gases, where the molecular population density is lower, the intensity is correspondingly lower and, in general, larger sample volumes must be used.

This state of affairs obviously calls for intense scources of monochromatic light and spectrographic instruments of high luminosity. Arc lamps emit numbers of intense lines, and, since each of these will excite its own Raman spectrum, there is the possibility of confusion due to overlapping. This can be reduced by the use of primary light filters chosen to isolate as far as possible one intense line of the source; however, due to the imperfect discrimination of most filters, it is clearly advantageous if the light source can be chosen so as to give an exciting line situated in a region which is as free as possible from other primary lines, especially on the low-frequency side where the Stokes Raman lines will appear. Much Raman work has been done with specially designed mercury-arc lamps, using the intense blue line (4358 Å), on the low-frequency side of which there is an almost clear range of about 2600 cm$^{-1}$. Since, as we shall see later, the intensity of the Raman effect (like that of Rayleigh scattering) is normally proportional to the fourth power of the frequency, the situation of Hg 4358 Å towards the high-frequency end of the visible is an advantage as regards intensity of scattering. The mercury-arc lamp also emits a strong line in the green (5461 Å) which has found application for samples which absorb in the blue region.

A chosen exciting line can be used in practice only if it (and the Raman lines to which it gives rise) is not seriously absorbed by the sample. Nor is a source practicable if it should give rise to serious fluorescence (which would mask the Raman spectrum) or cause photo-decomposition of the molecular species under investigation. On these accounts, light sources giving intense exciting lines in the yellow and red (and even in the near infrared) have been developed and used for "difficult" cases. In this connection, the helium discharge lamp deserves special mention. It gives intense lines at 5876 and 6678 Å and has been successfully used for the investigation of colored or photosensitive species for which blue (or green) mercury excitation is ruled out. For a general account see Stammreich.[4] Important also is the more recent application of lasers for the excitation of Raman spectra, which forms the subject of Chapter 3 of this book. Lasers have the advantage that they may generate only a single exciting line. Moreover, the frequency is highly monochromatic and the laser beam is emitted in a well-defined direction.

In general, in order to obtain very intense irradiation, the light source is constructed so as to surround the sample tube or the primary light is made to pass through it many times. The "aiming" of the spectrograph and the "coupling" of its collimator to the sample is generally an important practical matter. When it is remembered that the primary light intensity is so enormously greater than the Raman intensity, it becomes clear that the spectrograph should "see" only the sample and not any parts of the cell, *etc.* that might reflect unwanted primary light into the slit. Conditions for this "optical coupling" have been set forth by Nielsen,[5] and in practice care is taken to comply with them as far as possible. Only with a spectrographic instrument with very high discrimination between primary and Raman light (e.g., the commercially obtainable Cary Model 81 with its grating double monochromator, see Chapter 2) can less stringent coupling be used with advantage.

A very notable advance was achieved with the introduction[6] of the photomultiplier (with suitable amplifiers and pen recorder) in place of the photographic plate for the recording of Raman spectra. Although for the determination of line frequencies the photographic method is an excellent one, the photographic plate is a notoriously unsatisfactory agent for intensity measurements. Here photoelectric recording, with its linear response, is practically indispensable.

For a more detailed account of various aspects of experimental technique, reference should be made to Chapter 2 of this book.

### Applications in Chemical Analysis

It is not necessary to inquire into the theory of the Raman effect to appreciate its usefulness for "finger-print" applications. Each scattering species gives its own characteristic vibrational Raman spectrum, which can be used for its qualitative identification. In general, the spectrum is little affected by admixture with other species; and it is important to note that, since the spectrum usually consists of fairly sharp lines, it remains distinct and recognizable for purposes of qualitative analysis. Unlike infrared absorption, the Raman method is applicable without difficulty to aqueous solutions.

In systems where chemical interaction occurs, the presence of new molecular species can be detected by the appearance of new Raman lines. The method in no way disturbs chemical equilibrium and thus gives information about labile species which could not be apprehended by ordinary chemical analytical methods. Above all, because it gives distinct information about the individual species present, the Raman method is more powerful than colligative methods which measure only some property of the system as a whole. Hence, it is probably the most powerful means presently available for the study of ionic species and their equilibrium in concentrated aqueous solutions (see Chapter 7).

Since the intensity of a characteristic Raman line is to a fair approximation proportional to the volume concentration of the species in question, Raman intensity measurements provide a basis for quantitative analysis.

Useful as Raman spectra are for analytical purposes, however, it is necessary to look more closely at the theory of the Raman effect in order to be able to interpret individual spectra in terms of molecular properties.

## QUANTUM-MECHANICAL THEORY OF LIGHT SCATTERING

Soon after Raman's experimental discovery, theoretical treatments (classical, semi-classical, and quantum-mechanical) were developed. An authoritative article by Placzek,[7] published in 1934, gives (in addition to his own polarizability theory, see further below) a review of the fundamental quantum-mechanical theory associated with the names of Heisenburg, Kramers, Dirac, and others. No attempt will be made here to present the quantum-mechanical arguments: the form of the final expression for the intensity of a scattered line will be assumed without proof and briefly discussed. The treatment considers the perturbation of the wave-functions of the scattering molecule by the electric field of the incident light (frequency, $v_0$) and arrives at an expression for the induced electric moment matrix element $P_{nm}$ (otherwise known as the induced transition moment) associated with a transition between the initial state (designated by $n$) and the final state (designated by $m$) and defined by the following relation:

$$P_{nm} = \int \psi_m * P \psi_n \, d\tau$$

Here $P$ is the induced dipole moment, $\psi_m$ and $\psi_n$ are the time-independent wave functions of the states, and the integral is extended over the whole range of the coordinates.

The quantum-mechanical result is of the form

$$P_{nm} = \frac{1}{h} \sum_r \left( \frac{M_{nr} M_{rm}}{v_{rn} - v_0} + \frac{M_{nr} M_{rm}}{v_{rm} + v_0} \right) E \qquad (1)$$

where $h$ is Planck's constant, $r$ denotes any level of the complete set belonging to the unperturbed molecule, $v_{rn}$ and $v_{rm}$ are the frequencies corresponding to the differences between the states denoted by the subscripts, $M_{nr}$ and $M_{rm}$ are the corresponding transition moments, and $E$ is the electric vector of the incident light.

It is the square of $P_{nm}$ that determines the intensity of scattering of the line involving the induced transition $n \to m$, according to the equation

$$I_{nm} = C(v_0 + v_{nm})^4 P_{nm}^2 \qquad (2)$$

in which $C$ is a universal constant equal to $64\pi^2/(3c^2)$, where $c$ is the velocity of light. To obtain the total intensity from a system containing $N_n$ molecules in the initial state $n$, we must of course multiply by the factor $N_n$. The frequency $v_0 + v_{nm}$ is that of the scattered radiation. A form of equation (2) is taken as the starting point in Chapter 4 of this book.

When $n = m$ and hence $v_{nm} = 0$, the state of the scattering molecule remains unchanged and we have Rayleigh scattering. When $n \neq m$, we have the Raman effect. If the transition is to lower energy,

$$v_{nm} = (E_n - E_m)/h$$

is positive (anti-Stokes Raman line); if to higher energy, $v_{nm}$ is negative (Stokes line). We may note that it follows from the theory that Rayleigh scattering is coherent and Raman scattering incoherent. The expression for the intensity contains as a factor the fourth power of the frequency of the line in question.

The other factor $P_{nm}^2$ includes $E^2$ [see equation (1)], which shows the expected proportionality to the intensity of the incident light. As to the rest, we see that $P_{nm}$ involves a summation involving the transition moments from the initial state $n$ to all states $r$ of the unperturbed molecule and the transition moments from all states $r$ to the final state $m$. This does not mean, however, that such transitions in fact occur in the scattering act. The summation arises purely as a consequence of the mathematical treatment of the perturbation problem, in which a wave function of the perturbed molecule is expressed in terms of the full set of its unperturbed wave functions. It is important to note, in this connection, that the expression for $P_{nm}$ involves the *products* of transition moments with subscript $r$, and not the individual transition moments themselves. Since these quantities may be either positive or negative, it is possible for the terms belonging to different states $r$ either to reinforce one another or the reverse. They may indeed cancel out one another completely and so cause $P_{nm}$ to vanish. The Raman line is then forbidden.

The states $r$ are sometimes referred to as "intermediate levels" or "third common levels" of the scattering act; however, in adopting such nomenclature it must be borne in mind that their role is purely "*virtual*." The only necessary condition for Raman scattering is that the energy $hv_0$ of the incident photon must be greater than the energy difference $E_m - E_n$ between the final and initial states of the *actual* transition. Since no actual transition to any state $r$ takes place, $v_0$ is not matched to any absorption frequency of the scattering molecule. Indeed in general the summation we are discussing must include not only states $r$ which lie above the initial state $n$, but also states $r$ (if any)

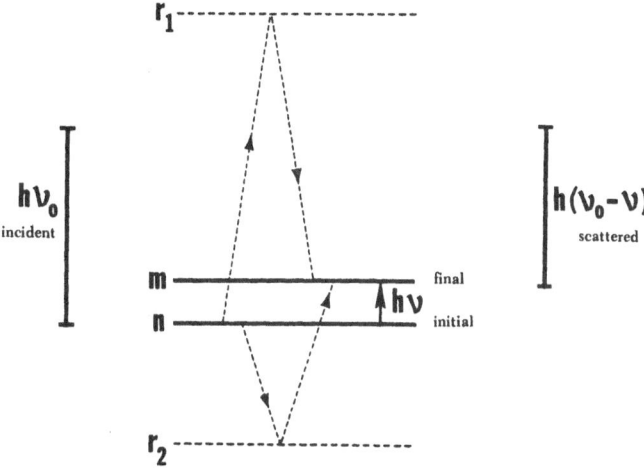

Fig. 1. Excitation of a Stokes Raman line. Actual transition $n \to m$ shown by a solid line with arrow. Virtual transitions involving two "third common levels" ($r_1$ above $n$ and $r_2$ below $n$) shown by broken lines.

which lie *below* it. In such cases, the idea of actually attaining the state $r$ by absorption of the incident light is obviously absurd.

In Fig. 1 the actual transition giving rise to a Stokes Raman line is indicated by a solid line with an arrow, and some typical virtual transitions are indicated by dotted lines.

From equations (1) and (2), we see that the dependence of the intensity of a Raman line upon the frequency $v_0$ of the exciting light is determined not only by the factor $(v_0 + v_{nm})^4$, but also by the denominators in the summation over the states $r$. In normal cases of colorless samples, $v_0$ is chosen in the visible region and so is small compared with both $v_{rn}$ and $v_{rm}$, the frequencies of electronic absorptions in the ultraviolet. Under these circumstances ($v_0 \ll v_{rn}$ or $v_{rm}$), the Raman intensity $I_{nm}$ simply becomes proportional to the fourth power of the frequency of the Raman line.

If, however, $v_0$ is chosen so as to lie near a particular absorption frequency $v_{rn}$, then the term containing the factor $1/(v_{rn} - v_0)$ will tend to become important in determining the intensity. In fact, $I_{nm}$ will rise markedly as $v_0$ approaches the absorption frequency. This is the so-called *resonance Raman effect*, which forms the subject of Chapter 6 of this book.

Equations (1) and (2), although of fundamental theoretical significance, are in practice of little use for the actual calculation of intensities, because the requisite knowledge of electronic levels and the associated transition moments is almost wholly lacking in most cases. Progress is possible, however, by theoretical treatments which introduce the polarizability of the molecule.

## POLARIZABILITY THEORY OF RAYLEIGH SCATTERING

### Tensor Character of the Polarizability

It will be convenient first to discuss Rayleigh scattering, and later (in the section that follows) to extend the discussion to the more complicated phenomenon of Raman scattering.

Attention is focused upon the electric moment $P$ induced in the scattering molecule by the electric field $E$ of the incident light. If $\alpha$ is the molecular polarizability, we may write

$$P = \alpha E \tag{3}$$

It should be clearly understood that we are here dealing only with the *induced* moment, which is quite distinct from any *permanent* moment which the molecule may possess in its unperturbed condition.

Since both $P$ and $E$ are vectors, $\alpha$ is in general a *tensor*. Only if the molecule is isotropic, in which case the directions of $P$ and $E$ are the same, does $\alpha$ become a scalar quantity. In the general, anisotropic case, let $E_x$, $E_y$, $E_z$ be the components of the electric vector relative to a space-fixed coordinate system, and consider first a fixed (nonrotating) molecule. The components of the induced moment are given by the following:

$$P_x = \alpha_{xx}E_x + \alpha_{xy}E_y + \alpha_{xz}E_z$$
$$P_y = \alpha_{yx}E_x + \alpha_{yy}E_y + \alpha_{yz}E_z \tag{4}$$
$$P_z = \alpha_{zx}E_x + \alpha_{zy}E_y + \alpha_{zz}E_z$$

Thus the tensor $\alpha$ is defined by an array of nine components $\alpha_{xx}$, $\alpha_{xy}$, *etc.* In all cases which interest us, it may be shown that the tensor is a symmetric one, i.e., the off-diagonal components are related by the following three equations:

$$\alpha_{xy} = \alpha_{yx} \qquad \alpha_{yz} = \alpha_{zy} \qquad \alpha_{zx} = \alpha_{xz} \tag{5}$$

The number of distinct components is thus reduced to six.

As a help in visualizing the tensor, we may make use of the six components to form the equation of an associated ellipsoid:

$$\alpha_{xx}x^2 + \alpha_{yy}y^2 + \alpha_{zz}z^2 + 2\alpha_{xy}xy + 2\alpha_{yz}yz + 2\alpha_{zx}zx = 1 \qquad (6)$$

It transpires that, if the coordinate system is rotated relative to the molecule, this equation transforms in the same way as the tensor itself. Hence, the ellipsoid may be regarded as representing the polarizability. In particular, if the coordinate axes are made to coincide with the principal axes $X$, $Y$, $Z$ of the ellipsoid, the equation reduces to the form

$$AX^2 + BY^2 + CZ^2 = 1 \qquad (7)$$

Relative to these particular axes, all the off-diagonal components of the tensor vanish. The diagonal components $\alpha_{XX} = A$, $\alpha_{YY} = B$, and $\alpha_{ZZ} = C$ are called the principal values of $\alpha$.

### Classical Treatment of Rayleigh Scattering: Degree of Depolarization

In Rayleigh scattering $\alpha$ is regarded as having no time dependence. The induced electric moment $P$ will thus oscillate with the same frequency $\nu_0$ as the incident light, and so will radiate scattered light which also has this frequency. This classical treatment may also be extended to give an account of the consequent state of polarization of the Rayleigh light. Imagine the molecule fixed at the origin of a

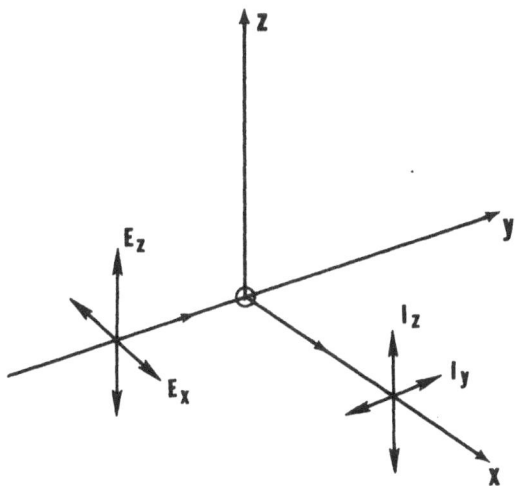

Fig. 2. Polarization of Rayleigh scattering for natural incident light.

space-fixed coordinate system and let it be irradiated with natural (i.e., unpolarized) light along the positive $y$-direction (Fig. 2). Consider now the scattering in the positive $x$-direction, i.e., at right angles to the incident direction. The incident light may be considered as consisting of two electric vectors $E_x$ and $E_z$ of equal magnitude. In general, each will induce oscillating electric moments. Each of the resulting components $P_x$, $P_y$, and $P_z$ will radiate Rayleigh light; however, since an oscillating moment cannot radiate in its own direction, only $P_y$ and $P_z$ will contribute to the scattering along the $x$-axis. The contributions will be plane-polarized with their electric fields in the $y$- and $z$-directions, respectively. Let their intensities (proportional to the squares of the radiating moments) be $I_y$ and $I_z$, respectively (Fig. 2). The sum of $I_y$ and $I_z$ is of course the total intensity $I_n$.

By equation (4), we have

$$P_y = \alpha_{yx}E_x + \alpha_{yz}E_z \tag{8a}$$

$$P_z = \alpha_{zx}E_x + \alpha_{zz}E_z \tag{8b}$$

In general, $P_y \neq P_z$ and so $I_y \neq I_z$. The ratio $I_y/I_z$ is known as the *degree of depolarization* of the scattered light and is denoted by $\rho_n$. The subscript $n$ here denotes that the case of *natural* (i.e., unpolarized) incident light is being considered. It should also be noted that, by convention, $\rho$-values always imply scattering *at right angles* to the incident direction.

In the special case where the scattering molecule is isotropic, $\alpha_{xy}$, $\alpha_{yz}$, and $\alpha_{zx}$ are all zero. Otherwise expressed, the polarizability ellipsoid is a sphere. When this is so, $P_y$ and $I_y$ vanish and $\rho_n$ is necessarily zero. The Rayleigh light is then said to be *completely polarized*. This conclusion is clearly independent of the orientation of the scattering molecule.

In fluids (as contrasted with single crystals), the orientation of the molecule will not be fixed, since the molecule will be rotating. In this case, the observed Rayleigh scattering from a sample will correspond to the average over all molecular orientations. The result is expressible in terms of two quantities associated with the tensor $\alpha$, which are called the *mean value* $\bar{\alpha}$ and the anisotropy $\gamma$ and which are defined as follows:

$$\bar{\alpha} = \tfrac{1}{3}(\alpha_{xx} + \alpha_{yy} + \alpha_{zz}) \tag{9}$$

$$\gamma^2 = \tfrac{1}{2}[(\alpha_{xx} - \alpha_{yy})^2 + (\alpha_{yy} - \alpha_{zz})^2 + (\alpha_{zz} - \alpha_{xx})^2 + 6(\alpha_{xy}^2 + \alpha_{yz}^2 + \alpha_{zx}^2)] \tag{10}$$

Both $\bar{\alpha}$ and $\gamma$ have the important property of being so-called *invariants*, i.e., their values are unaffected by changes of the orientation of the molecule relative to the space-fixed coordinate system. For averaging over all such orientations, we require the corresponding average values of the squares of the components of the tensor. These may be shown to be

$$\overline{(\alpha_{xx})^2} = \overline{(\alpha_{yy})^2} = \overline{(\alpha_{zz})^2} = \tfrac{1}{45}(45\bar{\alpha}^2 + 4\gamma^2) \tag{11}$$

$$\overline{(\alpha_{xy})^2} = \overline{(\alpha_{yz})^2} = \overline{(\alpha_{zx})^2} = \tfrac{1}{15}\gamma^2 \tag{12}$$

We are now in a position to write expressions for the intensities $I_y$ and $I_z$ of the plane-polarized components of the Rayleigh light (see Fig. 2) for freely rotating molecules. Referring to equation (8a), we see that $I_y$ has contributions proportional to $\overline{(\alpha_{xy})^2} \cdot E_x^2$ and to $\overline{(\alpha_{yz})^2} \cdot E_z^2$. Since $E_x = E_z = E$, we obtain

$$I_y = \text{const} \left(\tfrac{1}{15}\gamma^2 + \tfrac{1}{15}\gamma^2\right) E^2 \tag{13}$$

Similarly, we obtain from equation (8b),

$$I_z = \text{const} \left[\tfrac{1}{15}\gamma^2 + \tfrac{1}{45}(45\bar{\alpha}^2 + 4\gamma^2)\right] E^2 \tag{14}$$

Hence, for the total intensity $I_n$, it follows that

$$I_n = I_y + I_z = \text{const} \left[\tfrac{1}{45}(45\bar{\alpha}^2 + 13\gamma^2)\right] E^2 \tag{15}$$

and for the degree of depolarization

$$\rho_n = \frac{I_y}{I_z} = \frac{6\gamma^2}{45\bar{\alpha}^2 + 7\gamma^2} \tag{16}$$

For the special case of an isotropic molecule, $\gamma = 0$ and consequently $\rho_n = 0$, i.e., the Rayleigh line is completely polarized—a result we have noted previously. In general, since $\bar{\alpha}$ must clearly always be greater than zero, it follows that $\rho_n$ must always be less than $\tfrac{6}{7}$. Also, the intensity $I_n$ can never vanish, i.e., Rayleigh scattering can never be forbidden. It is produced not only by molecules, but also by atoms. (The corresponding conclusions are quite different for the case of Raman scattering.)

Incidentally, we may mention that considerations exactly similar to the above can be applied for the case where the incident light is *plane-polarized*, for example, with its electric vector in the $z$-direction $(E_x = 0)$. The resulting values for $I$ and $\rho$, conventionally written with the subscript $p$, are easily found to be

$$I_p = \text{const} \tfrac{1}{45}(45\bar{\alpha}^2 + 7\gamma^2)E^2 \qquad (17)$$

$$\rho_p = \frac{3\gamma^2}{45\bar{\alpha}^2 + 4\gamma^2} \qquad (18)$$

It is worth repeating that all conventional $\rho$-values imply scattering at right angles to the direction of the incident light. If the scattering angle is different from 90°, the value of $\rho$ is different—a circumstance that obviously gives rise to difficulties in experimental measurements, where precise definition of the angle must lead to serious loss of intensity.

The above classically derived results for Rayleigh scattering have been outlined in order to provide a simple starting point for the consideration (which now follows) of the more complicated phenomenon of Raman scattering, in which a quantum transition of the molecule is involved.

## GENERAL POLARIZABILITY THEORY OF RAMAN SCATTERING

### Range of Applicability

Placzek[7] has discussed the conditions under which the polarizability theory can be applied. The first is that the electronic state of the scattering molecule (in practice the ground state) shall be non-degenerate. This is fulfilled in all but a small minority of cases; although the theory may be modified to include them, the special phenomena which are then anticipated (see, for example, the work of Child and Longuet-Higgins[8]) will not be discussed here. We may note that in the above account of the polarizability theory of Rayleigh scattering it was tacitly assumed that electronic degeneracy was absent.

Placzek points out that, owing to the relatively large masses of the atomic nuclei, Raman scattering is normally due practically entirely to the electrons. The transference of energy between the incident light and the rotational or vibrational motions of the nuclei is thus occasioned by the influence of these motions upon the polarizability of the "electron cloud." Provided that the incident light frequency satisfies certain conditions (and in normal practice this is the case) the value of the polarizability for any instantaneous positions of the moving nuclei is practically identical with the value for stationary nuclei in the same positions. For this to be so, the incident frequency $v_0$ must be large in comparison with the nuclear frequencies in question, so that only the electrons can "follow" and not the relatively massive nuclei. Bearing in mind that the nuclear rotational and vibrational frequencies lie in the infrared, we see that this condition is obviously well satisfied when (as is usual) $v_0$ is in the visible region. A further condition is that the frequency difference $v_{rn} - v_0$ between an electronic absorption band and the incident light [which occurs as the resonance denominator in the quantum-mechanical equation (1)] shall also be large compared with the nuclear frequency. This condition is also well satisfied in practice, except in the extreme resonance case when $v_0$ approaches very close to an absorption frequency (see Chapter 6 of this book).

Within the limits of applicability thus defined, the polarizability $\alpha$ is a function only of nuclear positions. Hence, from a classical point of view, if the nuclei are performing a periodic motion of frequency $v_{mol}$, the polarizability must also have a time dependence of the same frequency.

It follows that, when the molecule is polarized by the electric vector of the incident light oscillating with frequency $v_0$, the "beat" frequencies $v_0 \pm v_{mol}$ will appear for the induced electric moment $P$. Thus, we have a classical explanation of anti-Stokes (positive sign) and Stokes (negative sign) Raman shifts. However, the classical theory is quite incapable of giving any account of the relative intensity of a Stokes line as compared with the corresponding anti-Stokes line.

In quantum-mechanical language it is necessary to consider, in place of the classical induced moment $P = \alpha E$, the matrix element $P_{nm}$ for the transition $n \rightarrow m$ in question. We write

$$P_{nm} = \left[\int \psi_m {}^* \alpha \psi_n \, d\tau\right] E = \alpha_{nm} \cdot E \qquad (19)$$

It will be appreciated that, in greater detail, expressions of this type must be used for each of the three components of the vector $P_{nm}$, taking account of the six components of the tensor $\alpha$ [compare equation (4)].

### Pure Rotational Raman Effect

In order to evaluate the matrix element $P_{nm}$ for pure rotational transitions, the components of the polarizability with respect to a space-fixed coordinate system are expressed in terms of the orientation of the molecule and the components with respect to coordinate axes fixed in the molecule. The calculation has been carried out by Placzek and Teller[9] and the selection rules established, *viz.*, all pure rotational transitions are forbidden in the Raman effect except those between pairs of levels of positive symmetry or those between pairs of levels of negative symmetry.

If the molecule is isotropic, no pure rotational Raman effect is possible. This is understandable classically; the polarizability ellipsoid is here a sphere, and as the molecule rotates it always presents the same polarizability aspect to the incident light. There is thus no modulation of $\alpha$ by the rotation, and no beat frequencies can arise. For a linear molecule, the selection rule is $\Delta J = \pm 2$. This also has a simple classical analog. By symmetry, one principal axis of the polarizability ellipsoid must be the line of the nuclei. As the molecule rotates about an axis at right angles to this line, the ellipsoid will assume an identical position after half a revolution and again, of course, after a complete revolution. The frequency of its change of aspect is thus twice the rotational frequency, and this corresponds to the quantum-mechanical selection rule $\Delta J = \pm 2$.

It also follows that only the anisotropy $\gamma$ (and not the mean polarizability $\bar{\alpha}$) can contribute to rotational scattering. In consequence [compare equation (16) for Rayleigh scattering], all pure rotational Raman lines must have the maximum possible value of $\rho_n$, namely, $\frac{6}{7}$. They are therefore said to be *fully depolarized*.

Of practical importance is the fact that the pure rotational Raman effect can be observed for molecules which possess no *permanent* electric depole moment and which, therefore, show no rotational absorption in the infrared or microwave regions. Thus, Raman spectroscopy can be used for the precise determination of molecular moments of inertia in cases where the absorption methods are wholly inapplicable. In practice, the pure rotational Raman shifts are small (except for a few very simple molecules with exceptionally low moments of inertia) and it is necessary to use a spectrograph of high dispersion and resolving power. For an authoritative account by one who has done extensive work in this field, the reader is referred to a review by Stoicheff.[10]

### Vibrational Raman Effect

A vibrational transition is in general accompanied by rotational transitions, so that the resulting Raman feature is in fact a vibration–rotation band. With gaseous samples and molecules of low moment of inertia, the rotational structure may be resolvable; however, with liquids this is no longer the case. The very great majority of Raman spectroscopic observations have been upon liquid samples, and so it is important to note that both classical and quantum-mechanical considerations lead to the conclusion that the total intensity of an unresolved vibration–rotational feature is identical with the intensity calculated for the pure vibrational transition when averaged over all orientations of the (nonrotating) molecule. We need therefore consider only vibrational transition moments of the form

$$P_{nm} = \left( \int \psi_m {}^* \alpha \psi_n \, dQ \right) E \tag{20}$$

where $\psi_m$ and $\psi_n$ are the vibrational wave functions and $\alpha$ is regarded as a function only of the nuclear *vibrational* coordinate $Q$.

In order to discuss the evaluation of $P_{nm}$ and hence the vibrational intensities, selection rules, and degrees of depolarization, we must first say something of the symmetry properties of molecules and their vibrational modes.

### Normal Vibrational Coordinates and their Symmetry Properties

In general the assemblage of nuclei of a molecule in its equilibrium configuration will possess certain *symmetry elements*, such as reflection planes, rotational axes, *etc.* With each such element is associated a *symmetry operation* (reflection in a plane, rotation through an appropriate angle about an axis, *etc.*), the performance of which produces a nuclear configuration indistinguishable from the original. The complete set of such operations for any molecule constitutes a *mathematical group*. Space will not permit an exposition of group theory,[11] but an example will be taken in order to introduce the reader to the general ideas and the nomenclature which is customarily used.

By virtue of its set of symmetry properties, a molecule can be assigned to a so-called *point group*, which is designated by a conventional symbol. Thus, for example, the planar $BF_3$ molecule (which

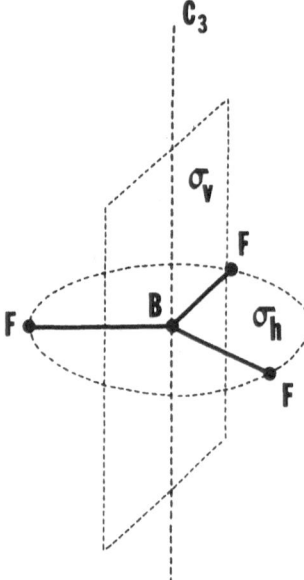

Fig. 3. Some symmetry properties
of the planar $BF_3$ molecule.

will serve to illustrate all the principal features) possesses one three-fold axis (designated as $C_3$ in Fig. 3) passing through the B atom and perpendicular to the $BF_3$ plane. Such an axis is conventionally taken as the vertical $z$-axis. The corresponding symmetry operations (which transform the equilibrium nuclear configuration into one which is indistinguishable from it) are rotation about the axis through $120°$ or $240°$. Rather confusingly, perhaps, these operations are also designated by the symbol $C_3$, usually with the distinguishing super-scripts 1 and 2. We may note that rotation through $360°$ brings the molecule back to the identical original configuration and so is wholly equivalent to the so-called identity operation $E$, which is necessarily always included in a group. In addition it is seen that the $BF_3$ molecule possesses a horizontal plane of symmetry (the plane of the molecule itself); it is designated by the symbol $\sigma_h$. There are also three vertical planes of symmetry ($\sigma_v$), each passing through the $C_3$ axis and one of the F nuclei. In consequence of all this the molecule is said to belong to the point group $D_{3h}$. It should be mentioned that, by virtue of the symmetry elements already enumerated, the $BF_3$ molecule also possesses others, $viz.$, three twofold axes $C_2$ (one along each of the

B–F bonds) and a threefold rotation–reflection axis $S_3$ which is coincident with $C_3$. The symmetry operations for the last-named are rotations as for $C_3$ followed by reflection in $\sigma_h$.

It is convenient here to mention certain necessary properties of a mathematical group (e.g., a group of symmetry elements) to which we shall have occasion to refer later. One is that the successive performance of any two group operations $R_1$ and $R_2$ is necessarily equivalent to some single operation $R_3$ of the group. This is expressed symbolically as $R_2 R_1 = R_3$ and it is conventional to refer to the successive performance $R_2 R_1$ as a "product" resulting from the "multiplication" of $R_1$ by $R_2$. Another property is that for each element $R$ there must be another element $S$ of the group (called the "reciprocal" of $R$) such that $RS = E$, where $E$ is the trivial identity element.

We come now to molecular vibrations. When the nuclear configuration is distorted slightly in any arbitrary way, restoring forces will come into play and, on being released, the molecule will perform vibrations of small amplitude. In general, the motion of the nuclei will appear to be random and aperiodic; however, analysis by classical mechanics shows that the seemingly complex motion is in fact compounded of a definite number of so-called *normal modes*, in each of which all the nuclei move with the same characteristic *normal vibrational frequency* and all in phase with one another.

The number of these normal modes in any particular case is easily deduced. If the number of nuclei is $N$, the total number of degrees of freedom (i.e., the number of coordinates required to specify the positions of all the nuclei) is $3N$. Of these, three account for translation of the molecule as a whole and three more for its rotation. This leaves $3N - 6$ to account for the *relative* positions of the nuclei, i.e., for internal motions. This then is the number of normal modes. For our example BF$_3$, the number is $(3 \times 4) - 6 = 6$. It should be noted that for a linear molecule, where one of the three rotations (that about the line of the nuclei) is impossible, the number of normal vibrational modes becomes $3N - 5$.

The symmetry of the equilibrium nuclear configuration (molecular point group) determines the symmetry of the "electron cloud" which holds the nuclei in their places. Hence, it determines also the symmetry of the restoring forces called into play by a small deformation. It will be appreciated, therefore, that it also determines the symmetry properties of the ensuing normal vibrational modes, each of which is

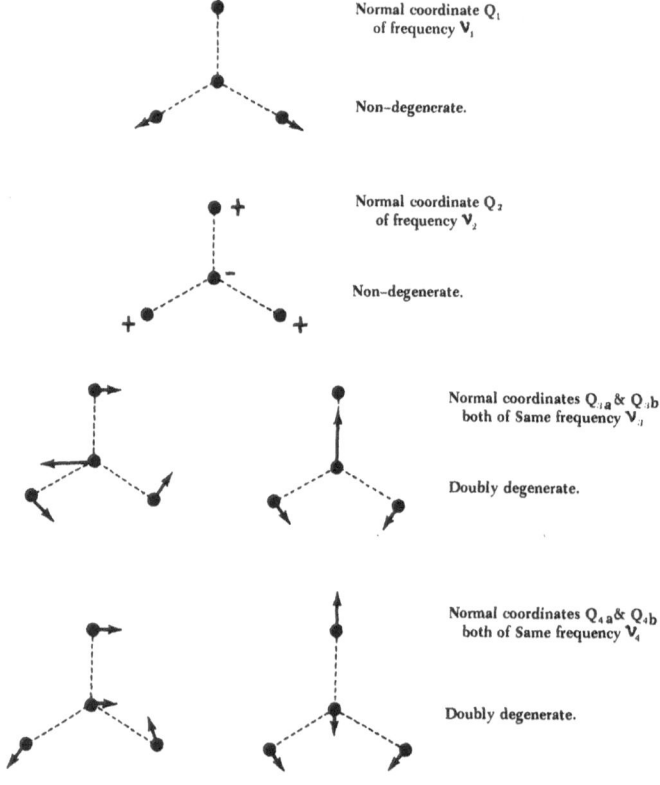

Fig. 4. Normal vibrational modes of a planar $MX_3$ molecule (e.g., $BF_3$). Arrows indicate directions and approximate relative amplitudes of nuclear displacements in one phase. For $Q_2$, a plus sign indicates displacement out of the plane of the molecule toward the observer and a minus sign indicates displacement away from the observer.

consequently said to belong to a certain so-called *symmetry species* (sometimes referred to as a symmetry class) according to its behavior under the symmetry operations of the group. Each mode is described by a corresponding *normal coordinate Q*, which expresses the appropriate displacements of all the nuclei.

For molecules possessing one or more axes of order greater than 2 (e.g., the $C_3$ axis of our example $BF_3$) it is found that, for symmetry reasons, certain pairs (or sets of three) of the normal modes necessarily have identical frequencies. These are then said to be *doubly* (or *triply*) *degenerate*. Examples are to be found in the normal modes of $BF_3$

which are indicated diagrammatically in Fig. 4. The frequencies $v_1$ and $v_2$ are *nondegenerate*, whereas $v_{3a}$ and $v_{3b}$ are equal and form a doubly degenerate pair. Similarly, $v_{4a}$ and $v_{4b}$ are also degenerate. The degeneracy arises because the rotational symmetry about the $z$-axis $(C_3)$ makes the $x$- and $y$-directions symmetrically equivalent. Hence, although the total number of normal modes is six (as deduced above), the number of *distinct frequency values* is only four.

### Symmetry Species and Group Character Tables

The symmetry species for the various possible point groups are conventionally presented in so-called *character tables* (see, for example, the work of Wilson et al.[12]) which usually take a form similar to that given in Table I for our example, the point group $D_{3h}$ to which the planar $BF_3$ molecule belongs.

The meaning of Table I will now be explained. In the top line, following the group symbol, are the symbols for all the *"classes" of symmetry operations* of the group. Note that the two rotations about the $C_3$ axis $(C_3^1$ and $C_3^2$, see above) are classed together as $2C_3$. Similarly the three $C_2$ rotations are classed together as $3C_2$, etc. The total number of symmetry operations is called the *order* $(g)$ *of the group*. The number of classes of symmetry operations is denoted by $r$. In our example, $g = 12$ and $r = 6$.

In order to understand the other entries in the character table it is necessary to remember that each symmetry operation, as performed upon a suitable set of nuclear coordinates, is equivalent to a linear transformation and so corresponds to an appropriate matrix. Any set of matrices, one for each operation, which behave with respect to multiplication in the same way as the members of the group is called a

### TABLE I
### Character Table for Point Group $D_{3h}$

| $D_{3h}$ | $E$ | $2C_3$ | $3C_2$ | $\sigma_h$ | $2S_3$ | $3\sigma_v$ | | |
|---|---|---|---|---|---|---|---|---|
| $a_1'$ | $+1$ | $+1$ | $+1$ | $+1$ | $+1$ | $+1$ | | $(\alpha_{xx} + \alpha_{yy}); \alpha_{zz}$ |
| $a_2'$ | $+1$ | $+1$ | $-1$ | $+1$ | $+1$ | $-1$ | $R_z$ | |
| $e'$ | $+2$ | $-1$ | $0$ | $+2$ | $-1$ | $0$ | $(T_x, T_y)$ | $(\alpha_{xx} - \alpha_{yy}); \alpha_{xy}$ |
| $a_1''$ | $+1$ | $+1$ | $+1$ | $-1$ | $-1$ | $-1$ | | |
| $a_2''$ | $+1$ | $+1$ | $-1$ | $-1$ | $-1$ | $+1$ | $T_z$ | |
| $e''$ | $+2$ | $-1$ | $0$ | $-2$ | $+1$ | $0$ | $(R_x, R_y)$ | $\alpha_{yz}; \alpha_{zx}$ |

*representation* of the group. Of particular interest is the set of matrices, each of size $3N \times 3N$, associated with transformations of the set of $3N$ Cartesian displacements of the $N$ nuclei. These coordinates, which are said to form the *basis* of the representation, will of course account not only for the $3N - 6$ vibrations but also for the three translations and the three rotations of the molecule as a whole. In general, the transformations which the matrices represent will lead to transformed coordinates which are linear combinations of the original set, i.e., will cause "mixing" of them. It may be shown that, when symmetry elements are present, the representation under consideration is *reducible* in the sense that, by a suitable choice of basis, it may be broken down into matrices of smaller size which cause mixing of fewer of the basis coordinates. When this breakdown process has been carried as far as possible, we obtain a certain number of so-called *irreducible representations*, such that the extent of mixing of basis coordinates has been reduced to a minimum. It may be shown that the number of different irreducible representations is equal to the number $r$ of classes of symmetry operations; in our example (Table I), this number is six. In size the matrices of a particular irreducible representation may be all $1 \times 1$ (i.e., pure numbers, and in fact all either $+1$ or $-1$); or all $2 \times 2$; or all $3 \times 3$. It transpires that the set of $3N - 6$ normal vibrational coordinates together with the three translations and the three rotations form a basis for the irreducible representations of the point group of a molecule. The different irreducible representations correspond to the *symmetry species*, to which reference has been made in the preceding section.

The normal coordinate of a mode belonging to a species whose irreducible representation is made up of $1 \times 1$ matrices (i.e., numbers) does not mix with any other and is thus nondegenerate. Normal coordinates belonging to a species whose irreducible representation is made up of $2 \times 2$ matrices occur in pairs which mix together and must have identical frequencies. They are thus doubly degenerate. Similar considerations apply to modes belonging to triply degenerate species.

In the character table (Table I), the first column lists the symbols for the symmetry species. Nondegenerate species are conventionally designated by the letters $a$ or $b$ (sometimes by the corresponding capital letters) bearing certain subscripts and superscripts. Doubly degenerate species are similarly designated by $e$ and triply degenerate by $f$.

One might expect that a descriptive table would contain the actual matrices of the different irreducible representations, and in fact this is the case where these are simple numbers (see Table I for the nondegenerate species $a'_1$, $a'_2$, $a''_1$ and $a''_2$). For degenerate species an economy of space is possible. For all practical purposes (see below) it is sufficient to tabulate, not the set of $2 \times 2$ matrices for an $e$ species or the set of $3 \times 3$ matrices for an $f$ species, but merely in each case the *sum of the diagonal elements* of each matrix. This sum, which is of course a simple number, is called the *character* $\chi$ of the matrix—hence the name character table. (We remark that for a $1 \times 1$ matrix, the matrix and its character are identical.) The grouping of symmetry operations (in our example, 12 of them) into classes (in our example, 6) allows further economy in the tabular presentation, since all the operations in a given class (e.g., the three $C_2$ operations in our example) necessarily have the same character.

Examination of the normal modes of $BF_3$ shown diagrammatically in Fig. 4 will verify that $Q_1$ belongs to the so-called *totally symmetric* species $a'_1$. This mode passes over into itself under all the operations of the group, i.e., the matrices of the irreducible representation or symmetry species are all simply $+1$ (see Table I). For $BF_3$ this is the sole normal mode of this particular species. Similarly, $Q_2$ is the sole mode of species $a''_2$.

Much more detailed examination shows that the degenerate pair $Q_{3a}$, $Q_{3b}$ (and similarly the pair $Q_{4a}$, $Q_{4b}$) belongs to the species $e'$. An idea of the sort of considerations involved here may be obtained by looking at the displacement vectors of the central nucleus in the coordinates $Q_{3a}$ and $Q_{3b}$. In Fig. 4 these are shown as equal in magnitude. As far as this nucleus alone is concerned, the effect of the operation $C_3^1$ (clockwise rotation through 120° about the $C_3$ axis) is to transform the displacement $Q_{3a}$ into $Q'_{3a}$ which is easily seen to be the vector sum

$$-Q_{3a}\cos 60° + Q_{3b}\cos 30° = -\tfrac{1}{2}Q_{3a} + \tfrac{\sqrt{3}}{2}Q_{3b}$$

(Note the "mixing" of the two normal coordinates.) Similarly the same operation transforms $Q_{3b}$ into

$$Q'_{3b} = -\tfrac{\sqrt{3}}{2}Q_{3a} - \tfrac{1}{2}Q_{3b}$$

The transformation matrix is thus

$$\begin{bmatrix} -\tfrac{1}{2} & +\tfrac{\sqrt{3}}{2} \\ -\tfrac{\sqrt{3}}{2} & -\tfrac{1}{2} \end{bmatrix}$$

It is found that exactly the same matrix applies for the transformations of the displacements of each of the other nuclei. It therefore applies to the normal coordinates as a whole. The character of the matrix is $-\frac{1}{2} -\frac{1}{2} = -1$. Thus, the corresponding entry in the character table (species $e'$, symmetry operation class $2C_3$) is verified (see Table I).

We have thus accounted for all six normal vibrational modes of $BF_3$. The three translations $T_x$, $T_y$, $T_z$ in the directions indicated by the subscripts, and likewise the three rotations $R_x$, $R_y$, $R_z$ about the respective Cartesian axes, also belong to symmetry species. Their allocation is shown in the penultimate column of the character table (q.v.). Note that $T_z$ and $R_z$ are nondegenerate, whereas the pairs $(T_x, T_y)$ and $(R_x, R_y)$ are each doubly degenerate. This is of course a consequence of the rotational symmetry about the $z$-axis $(C_3)$ which makes the $x$- and $y$-directions symmetrically equivalent.

The character table holds for all molecules belonging to the point group $D_{3h}$. For our special example $BF_3$, the 12 coordinates forming the basis of the completely reduced representation $\Gamma$ are allocated to the different symmetry species as shown in Table II.

This is expressed by writing

$$\Gamma = a_1' + a_2' + 3e' + 2a_2'' + e'' \tag{21}$$

By removing the transitions and rotations (see Table II) we obtain for the vibrations alone

$$\Gamma_{vib} = a_1' + 2e' + a_2'' \tag{22}$$

**Selection Rules**

As we have seen from equation (2), the intensity of a Raman transition $n \rightarrow m$ is proportional to the square of the transition moment

### TABLE II
#### Species of Translations, Rotations, and Vibrations of $BF_3$

| Species | Coordinates |
|---------|-------------|
| $a_1'$  | $Q_1$ |
| $a_2'$  | $R_z$ |
| $e'$    | $(T_x, T_y) : (Q_{3a}, Q_{3b}) : (Q_{4a}, Q_{4b})$ |
| $a_1''$ | — |
| $a_2''$ | $T_z ; Q_2$ |
| $e''$   | $(R_x, R_y)$ |

$P_{nm}$. For vibrational transitions, $P_{nm}$ is defined by equation (20). If the integral in this expression vanishes, the transition is forbidden. This is the basis of the selection rules.

Consider the vibrational *fundamental* of a single normal mode, i.e., the transition between the ground vibrational state (vibrational quantum number $v = 0$) and the state with $v = 1$. The time-independent wave functions $\psi_0$ and $\psi_1$, being functions of the normal coordinate $Q$, belong to symmetry species of the point group of the molecule. It may be shown that in all cases $\psi_0$ belongs to the totally symmetric species and that $\psi_1$ belongs to the same species as the vibration itself. (It is here assumed that the vibration is simple harmonic.)

The polarizability is also a function of the normal coordinate $Q$, and its six components likewise belong to symmetry species of the group. Their proper allocation is easily found, since a component $\alpha_{ij}$ is of the same species as the product of the translations $T_i$ and $T_j$.

The explanation of the character table (Table I) may now be completed. The last column shows the species of the components of the polarizability, or suitable linear combinations of them.

We are now in a position to derive the *selection rules*. These depend upon the fact that, bearing in mind that $\psi_0$ is totally symmetric, the integral in the expression for $P_{nm}$ [equation (20)] must vanish unless $\psi_1$ and $\alpha$ both belong to the same symmetry species. Hence, *a fundamental is permitted in the Raman effect only if at least one component of the polarizability is of the same species as the vibration itself.*

Inspection of the character table (Table I) and the species of the normal vibrations (Table II) will show at once that the fundamentals $v_1$, $v_3$ and $v_4$ of $BF_3$ are all permitted in the Raman effect, but that $v_2$ is forbidden. As far as fundamentals are concerned, therefore, the Raman spectrum will consist of three lines.

It is important to note here that the Raman selection rules are, in general, different from those for infrared absorption. In the latter, the operative transition moment is

$$\int \psi_m^* \mu \psi_n \, dQ$$

where $\mu$ is the *permanent* electric dipole moment of the molecule. The species of a component of the vector $\mu$ is the same as that of the corresponding translation, as given in the character table. A fundamental is permitted in the infrared only if at least one component of $\mu$ is of the same species as the vibration itself. The reader may confirm that, for our example $BF_3$, this leads to the result that $v_2$, $v_3$ and $v_4$ are permitted, but $v_1$ is forbidden—a state of affairs different from that in the Raman effect. This example emphasizes the complementary character of the two kinds of spectra, since in the case of $BF_3$, in order to observe all four fundamentals, *both* methods must be used. A similar state of affairs is found for many other molecules.

Although we have so far dealt only with fundamentals, exactly similar considerations can be applied to obtain the selection rules for overtones and combination tones. In the Raman effect, however, these have relatively low intensities and are only seldom observed. For this practical reason, a detailed discussion of them will be omitted here. (For such a discussion, see, for example, Herzberg's well-known book.[13])

## Rules of Polarization

We have seen that the degree of depolarization of the Rayleigh line is given by equation (16):

$$\rho_n = \frac{6\gamma^2}{45\bar{\alpha}^2 + 7\gamma^2} \tag{16}$$

in which $\bar{\alpha}$ and $\gamma$ are the mean-value and anisotropy invariants of the polarizability tensor $\alpha$. In the treatment of the Raman effect, the place of $\alpha$ is taken by

$$\alpha_{nm} = \int \psi_m^* \alpha \psi_n \, dQ$$

and we are led to an exactly analogous expression for the degree of depolarization $\rho_n$. The consequences may be exemplified for the fundamentals of $BF_3$. References to the character table (Table I) shows that the only components of $\alpha$ of the same species ($a_1'$) as $Q_1$ are $(\alpha_{xx} + \alpha_{yy})$ and $\alpha_{zz}$. This means that the relevant mean-value invariant will not vanish. In consequence [see equation (16) above] the Raman line of frequency $v_1$ will have $\rho_n < \frac{6}{7}$, i.e., will be *polarized*. On the other hand, for the pair $(Q_{3a}, Q_{3b})$, the only components of $\alpha$ of the same species ($e'$) are $(\alpha_{xx} - \alpha_{yy})$ and $\alpha_{xy}$. Since neither $(\alpha_{xx} + \alpha_{yy})$ nor $\alpha_{zz}$ occurs, the relevant mean-value invariant must vanish, and in consequence $\rho_n$ must have the value $\frac{6}{7}$, i.e., the Raman line of frequency $v_3$ must be *fully depolarized*. The same is true for $v_4$.

Considerations of this kind show, in general, that *all Raman lines of fundamentals belonging to the totally symmetric species are polarized*

$(\rho_n < \frac{6}{7})$ and that all Raman lines of other species are depolarized $(\rho_n = \frac{6}{7})$.

## Numbers of Fundamentals in Each Symmetry Species

In order to predict the Raman spectrum for any given molecular model, it is necessary to find the numbers of fundamentals belonging to each symmetry species, or (what is equivalent) the number of times $n_i$ that each irreducible representation $\Gamma_i$ appears in the completely reduced representation $\Gamma$. This can be done by making use of the character table.

Consider first the reducible $3N \times 3N$ representation whose basis is the set of $3N$ Cartesian displacements of the $N$ nuclei. For a given symmetry operation $R$, the only contributions to the character $\chi_R$ of the corresponding matrix (which, we recall, is the sum of the *diagonal* components) obviously come from nuclei which are not shifted by the operation. With a little thought, these contributions can be written down for any symmetry operation. For our point-group example $D_{3h}$, they are as shown in Table III. Also tabulated are the numbers of unshifted nuclei for the particular example $BF_3$ and the consequent value of $\chi_R$ for each symmetry operation.

Group theory shows that these $\chi_R$-values remain unchanged when the set of basis coordinates is transformed and, in particular, when it is transformed to the set of normal vibrational coordinates together with the translations and rotations. It is also true that $\chi_R$ is equal to the sum of the characters $\chi_{iR}$ for all the irreducible representations which occur in the completely reduced representation. Thus, if the $i$th irreducible representation occurs $n_i$ times, then

$$\chi_R = \sum_i (n_i \chi_{iR}) \tag{23}$$

## TABLE III
### Characters $\chi_R$ for Planar $BF_3$ Molecule

| Point group $D_{3h}$ | $E$ | $2C_3$ | $3C_2$ | $\sigma_h$ | $2S_3$ | $3\sigma_v$ |
|---|---|---|---|---|---|---|
| Contribution to $\chi_R$ per unshifted nucleus | +3 | 0 | −1 | +1 | −2 | +1 |
| Number of unshifted nuclei for $BF_3$ | 4 | 1 | 2 | 4 | 1 | 2 |
| $\chi_R$ for $BF_3$ | +12 | 0 | −2 | +4 | −2 | +2 |

We thus have a set of equations (one for each class of symmetry operations) and we may solve them to obtain the unknowns $n_i$. The solution is

$$n_i = \frac{1}{g} \sum_R (\chi_{iR} \chi_R) \tag{24}$$

where $g$ is the total number of symmetry operations, i.e., the order of the group.

As an example, we may again take the $BF_3$ molecule, for which the $\chi_{iR}$ values are given in the character table (Table I) and the $\chi_R$ values in Table III. The order $g$ of the group $D_{3h}$ is 12. For the $a_1'$ species, for instance, by applying equation (24) we find

$$n_{a_1'} = \tfrac{1}{12}[(1 \times 12) + 2(1 \times 0) + 3(1 \times -2) + (1 \times 4) + 2(1 \times -2)$$
$$+ 3(1 \times 2)]$$
$$n_{a_1'} = 1$$

As there is no translation or rotation of this species (see Table II), there must be just one normal vibration. This is the one of frequency $v_1$, in agreement with Fig. 4. For the $a_2''$ species, we find

$$n_{a_2''} = \tfrac{1}{12}[(1 \times 12) + 2(1 \times 0) + 3(-1 \times -2) + (-1 \times 4)$$
$$+ 2(-1 \times -2) + 3(1 \times 2)]$$
$$n_{a_2''} = 2$$

One of the two is accounted for by the translation $T_z$ (see Table II) and so there must be just one normal vibration of this species. It is the one of frequency $v_2$ (see Fig. 4). In exactly the same way we find for the species $e'$ that $n_{e'} = 3$. One degenerate pair is $(T_x, T_y)$ and so there must be two degenerate pairs of normal vibrations. They are the pairs with frequencies $v_3$ and $v_4$, respectively (see Fig. 4). This accounts for all six normal vibrations, and similar calculations will of course confirm that there are no more in any of the other symmetry species.

## Experimental Determination of Molecular Symmetry

The reader will now appreciate how, from the Raman spectrum, it is possible in principle to discriminate between alternative models that might be proposed for a given molecular species. Suppose, for example, that we wish to find out whether a molecule $MX_3$ (in which all three X atoms are attached to a central M atom) is planar like $BF_3$ or pyramidal like $PF_3$. The respective point groups are $D_{3h}$ and $C_{3v}$. In both cases, of course, the total number of normal vibrations is six. For the planar model [as we have already seen, equation (22)]

$$\Gamma_{vib} = a_1' + a_2'' + 2e'$$

In the same way, we obtain for the pyramidal model

$$\Gamma_{vib} = 2a_1 + 2e$$

i.e., two modes belong to the totally symmetric species $a_1$ and two pairs to the doubly degenerate species $e$. Thus, there are again (as for the planar model) four distinct fundamental frequencies; however, the rules of selection and polarization in the Raman effect are different. Reference to the appropriate character table (point group $C_{3v}$) shows that for the pyramidal model (unlike the planar one) all four fundamentals are permitted. Moreover they are all also permitted in infrared absorption. Of the Raman lines, the two of species $a_1$ will be polarized and the two of species $e$ depolarized.

The predictions for the two models are shown in Table IV (*cf.* Chapter 7, where similar results are tabulated for a large number of molecular types).

It is at once clear from Table IV that from the Raman spectrum alone it is in principle possible to decide which of the two symmetries a molecule $MX_3$ possesses. For instance, if two Raman lines are found to be polarized, or if all four fundamentals appear, then the molecule must be pyramidal. Very valuable confirmation can be obtained from the infrared absorption spectrum. Thus, if the molecule is planar, the single Raman frequency $v_1$ which is polarized is forbidden in the infrared, whereas if it is pyramidal both the polarized Raman lines will be permitted in the infrared. It is important to realize that all conclusions

### TABLE IV
### Predictions for Planar and Pyramidal $MX_3$ Models

| | Distinct frequencies | Permitted in Raman effect | Polarized Raman lines | Permitted in infrared absorption | Coincident frequencies in Raman and infrared |
|---|---|---|---|---|---|
| Planar ($D_{3h}$) | 4 | 3 | 1 | 3 | 2 |
| Pyramidal ($C_{3v}$) | 4 | 4 | 2 | 4 | 4 |

of this kind are *based solely upon symmetry considerations* and require no knowledge of force fields.*

The method is often very powerful. It must, however, be mentioned that in practice a number of circumstances may detract from its usefulness. We have tacitly assumed that the complete number of Raman-active (and infrared-active) fundamentals has been observed. It may happen, however, that certain of them are of such low intensity as to have escaped observation. Also, a polarized Raman line ($\rho_n < \frac{6}{7}$) may have a $\rho_n$-value which is so near $\frac{6}{7}$ that its polarized character may not have been recognized. There is also one circumstance that may occasionally make the number of observed Raman lines greater than would have been expected. This may occur if a permitted fundamental frequency happens fortuitously to be very nearly equal to an overtone or combination frequency *which is of the same species*, when the two may enter into so-called *Fermi resonance*. In consequence, the frequencies are somewhat displaced and the overtone or combination (which would otherwise have been relatively very weak) may "borrow" intensity from the fundamental. Thus, two lines may appear which might both be mistaken for fundamentals. The classic example occurs for $CO_2$ (linear), for which the rules permit only one (polarized) Raman line ($v_1$). By chance, $v_1$ is here nearly equal to the overtone $2v_2$ of the same species and is in Fermi resonance with it. Consequently, instead of the single expected Raman line, a Fermi doublet appears.

Insofar as infrared absorption observations are adduced for the determination of molecular symmetry, it must be borne in mind that corresponding limitations also apply in infrared spectroscopy.

## PLACZEK'S SIMPLIFIED POLARIZABILITY THEORY

### Taylor-Series Expansion of Polarizability

In the general polarizability theory outlined above, the tensor $\alpha$ has been regarded as a function of the nuclear coordinates. Thus, for a particular vibrational mode, $\alpha$ is a function of the normal coordinate $Q$. Bearing in mind that the amplitude of $Q$ is small, we may follow Placzek[7] and expand $\alpha$ as a Taylor series breaking off after the linear term, i.e., neglecting higher powers of $Q$:

* See the section, "Calculation of Force Fields from Frequencies," pp. 32–35.

$$\alpha = \alpha_0 + \left(\frac{\partial \alpha}{\partial Q}\right)_0 Q \qquad (25)$$

Here $(\partial\alpha/\partial Q)_0$ is the rate of change of $\alpha$ with respect to $Q$ at the equilibrium configuration (subscript zero) of the molecule. In more detail, an expression of the above form holds for each of the six distinct components of $\alpha$.

The corresponding matrix element for the transition $n \to m$ becomes

$$\alpha_{nm} = \int \psi_m^* \alpha_0 \psi_n \, dQ + \left(\frac{\partial \alpha}{\partial Q}\right)_0 \int \psi_m^* Q \psi_n \, dQ \qquad (26)$$

Because of the mutual orthogonality of $\psi_m$ and $\psi_n$, the first term on the right vanishes unless $n = m$. It thus corresponds to Rayleigh scattering. If simple harmonic vibration is assumed, it is a known property of the wave functions that the second term vanishes unless $m = n \pm 1$. It thus corresponds to the Raman scattering of fundamentals. According to this simplified theory, overtones and combination tones are forbidden: this is a consequence of the neglect of higher terms in the Taylor expansion. Actually, overtones and combinations only appear very weakly, if at all, so that the approximation is well justified in practice.

Consider the Stokes line of the fundamental of a particular mode, i.e., transition $n \to n + 1$, where the molecule is nonrotating. With insertion of the known value[14] of the integral, the expression for $\alpha_{n,n+1}$ becomes

$$\alpha_{n,n+1} = \left(\frac{\partial \alpha}{\partial Q}\right)_0 \sqrt{\frac{(n+1)h}{8\pi^2 M v}} \qquad (27)$$

in which $M$ and $v$ are the reduced mass and the fundamental frequency of the mode, respectively. The intensity and state of polarization of the Raman line depend upon the so-called *derived tensor* $(\partial\alpha/\partial Q)_0$, which we will designate by $\alpha'$. This is exactly analogous to the actual polarizability tensor $\alpha$: in fact, each component of $\alpha'$ is the derivative with respect to $Q$ of the corresponding component of $\alpha$. Like $\alpha$, the derived tensor possesses invariants, which we will write as $\bar\alpha'$ and $\gamma'$. Considerations exactly similar to those for Rayleigh scattering*

---

* See the subsection, "Classical Treatment of Rayleigh Scattering: Degree of Depolarization," pp. 11–14.

lead to the result that for an assemblage of rotating molecules irradiated with natural light, the total intensity $I_n$ and the degree of depolarization $\rho_n$ of the Raman light scattered at right angles to the direction of incidence are given by

$$I_n = \text{const } \tfrac{1}{45}[45(\bar{\alpha}')^2 + 13(\gamma')^2] \tag{28}$$

and

$$\rho_n = \frac{6(\gamma')^2}{45(\bar{\alpha}')^2 + 7(\gamma')^2} \tag{29}$$

New features arise, however, as compared with Rayleigh scattering. This is because the components of the derived tensor (being derivatives with respect to $Q$) may be either positive, negative, or zero. In consequence, not only $\gamma'$ but also $\bar{\alpha}'$ may vanish. (For the tensor $\alpha$, of course, $\bar{\alpha}$ can never vanish.) When both $\bar{\alpha}'$ and $\gamma'$ are zero, the Raman line is forbidden. (We have seen, in contrast, that Rayleigh scattering can never be forbidden). Moreover, when $\bar{\alpha}'$ alone vanishes, the degree of depolarization of the Raman line must have the maximum value $\tfrac{6}{7}$. (We have seen that for Rayleigh scattering $\rho_n$ must always be less than $\tfrac{6}{7}$.)

The rules of selection and polarization have already been discussed in terms of the general polarizability theory,[*] and so we need not consider them again at length in terms of Placzek's more restricted theory based upon the two-term Taylor expansion of $\alpha$ [equation (25)]. From equation (27) we see that, confining ourselves to simple-harmonic fundamentals, the derived tensor $\alpha'$ plays the role in the limited theory which is played by the tensor $\alpha_{nm}$ in the general theory. Thus, symmetry considerations which cause one or both of the invariants of $\alpha_{nm}$ to vanish will produce the corresponding results for the derived tensor $\alpha'$. This confirms that the rules of selection and polarization for simple-harmonic fundamentals are the same as those deduced from the general theory.

## CALCULATION OF FORCE FIELDS FROM FREQUENCIES

### Calculation of Normal Frequencies and Modes

The normal vibrational frequencies and modes of a molecule are determined by the masses and configuration of the nuclei and by

---

[*] See subsections, "Selection Rules" and "Rules of Polarization," pp. 24–27.

the restoring forces which come into play when the nuclear configuration is distorted. *If the force field is known or assumed*, the frequencies
and modes may be calculated by classical mechanics. No attempt
will be made here to expound the nature of this calculation in detail.
For a full treatment the reader is referred to the authoritative book
by Wilson, Decius, and Cross.[15] Some general comments are called
for, however, in relation to the subsequent discussion of Raman
intensities of this chapter.

To solve the equations of classical vibrational motion, it is
necessary to set down expressions for the potential and kinetic energies
in terms of some set of $3N - 6$ internal displacement coordinates.
In the harmonic approximation, which is always assumed in practice,
the potential energy is the sum of quadratic terms whose coefficients
are the *force constants*. Physical considerations make it appropriate
to choose internal displacement coordinates involving changes of
bond lengths and bond angles; the problem assumes its most elegant
and tractable form when molecular symmetry is taken fully into account
and the coordinates chosen are the so-called *symmetry coordinates*,
i.e., linear combinations of the internal coordinates constructed so
as to belong (like the desired normal coordinates) to the symmetry
species of the point group in question. The potential and kinetic
energy expressions may be used to derive the so-called *secular equation*,
the solution of which gives the normal frequencies. At the same time,
the *normal coordinates* corresponding to the frequencies may be found
as linear combinations of the symmetry coordinates, giving a description of each normal mode in terms of the directions and relative
amplitudes of displacement of each of the nuclei (compare Fig. 4).

### Calculation of Force Fields from Frequencies

In practice the force field is not known in advance, and the object
is to reverse the procedure described in the preceding section and to
calculate the force constants from fundamental frequencies that
have been observed in Raman and infrared spectra. For this purpose
it is necessary first to have assigned the observed frequencies to their
proper symmetry species. Some guidance in doing this is given by
spectroscopic evidence (e.g., polarization of Raman lines), but often
this is insufficient to determine the assignments unambiguously.
The calculation itself is not straightforward. The procedure that has
to be adopted is to start off with some plausible trial force field and to

calculate a set of frequency values from it. These will not agree with the observed values, but a comparison of the two sets provides a basis for improving the trial force constants. This process of improvement is repeated successively until (in favorable cases) a reasonably satisfactory fit between calculated and observed frequencies is attained.

In all this there is, however, a difficulty of principle, namely, that the number of force-constant parameters required for a full description of the field is in general greater (and often considerably greater) than the number of observable frequencies. Thus, the frequency data are insufficient to determine the force field completely. In practice it is therefore necessary to work with *simplified fields* requiring fewer force-constant parameters for their description. In fact, the number of force constants must not exceed (and, to allow for tests of the field, should be less than) the number of frequencies available. The choice of such simplified fields must be based on physical intuition, and this unfortunately introduces a measure of arbitrariness into the whole undertaking.

The process of successive modification of an assumed set of force-constant parameters so as to produce successively closer fits of calculated and observed frequencies is sometimes beset by special difficulties and is at best a laborious one. It may, however, be programmed for an electronic computer, and thus a "best-fit" set of force constants, based upon a least-squares frequency-fit criterion, may be rapidly obtained. At the same time, the computation will give the corresponding normal coordinates describing the nuclear motions (stretching of bonds, changing of bond angles, *etc.*) in each normal mode.

It must not be forgotten, however, that the force field so derived is only the best that can be done with the particular type of approximate, simplified field assumed at the beginning of the calculation. Indeed, physically different guesses as to field type may give equally good "best fits;" none of them will correspond to reality, even in the simple harmonic approximation. Important for our purposes is the fact that the calculated normal coordinates describing the normal modes will also be only approximations. Nevertheless, despite the uncertainties involved, the results obtained are not without interest provided that the type of force field used is a physically plausible one.

In suitable cases, the situation can be improved by increasing the number of observable frequencies by isotopic substitution in the molecule (especially replacement of *H* by *D*) which leaves the force field unchanged, and also by including in the force-field calculation other experimentally accessible data, such as Coriolis coupling constants, vibrational amplitudes, *etc.*

## RAMAN INTENSITIES ON THE BASIS OF PLACZEK'S THEORY

Consider the Stokes Raman line of a particular vibrational fundamental $(n \to n + 1)$. With use of equation (27), the induced transition moment is given by

$$P_{n,n+1} = (\alpha_{n,n+1})E = \alpha'_0 \sqrt{\frac{(n+1)h}{8\pi^2 M v}} \qquad (30)$$

In detail, an expression of this sort applies to each component of $P_{n,n+1}$. Intensity is proportional to the square of the transition moment

$$I = \text{const} \, (v_0 - v)^4 (P_{n,n+1})^2 \qquad (31)$$

where $(v_0 - v)$ is the frequency of the Stokes fundamental Raman line. Hence, for irradiation with natural light and for scattering at right angles, we obtain [*cf.*, equation (28)]

$$I_n = \text{const} \, (v_0 - v)^4 \frac{(n+1)h}{8\pi^2 M v} \cdot \frac{E^2}{45} [45(\bar{\alpha}')^2 + 13(\gamma')^2] \qquad (32)$$

At ordinary temperatures most of the scattering molecules in a sample will be in the lowest vibrational level $(n = 0)$, but some will be in higher levels for which the intensity (proportional to $n + 1$) will be higher. This must be taken into account in calculating the actual intensity which will be observed. According to the Boltzmann distribution law, the fraction $f_n$ of molecules with vibrational quantum number $n$ is

$$f_n = \frac{e^{-(n+\frac{1}{2})hv/kT}}{\sum_n e^{-(n+\frac{1}{2})hv/kT}} \qquad (33)$$

The total intensity will be proportional to $\sum_n f_n(n + 1)$, which may easily be shown to be equal to $(1 - e^{-hv/kT})^{-1}$. Using equation (32), we therefore obtain for the observed intensity

$$(I_n)_{\text{obs}} = K I_0 \frac{(v_0 - v)^4}{M v (1 - e^{-hv/kT})} [45(\bar{\alpha}')^2 + 13(\gamma')^2] \qquad (34)$$

in which $I_0$ (proportional to $E^2$) is the incident intensity and $K$ is a constant. Note that this result applies for irradiation with natural light. If the incident light is plane-polarized with its electric vector at right angles to the direction of scattering (i.e., in the $z$-direction in Fig. 2), the corresponding result [cf., equation (17)] is

$$(I_p)_{obs} = KI_0 \frac{(v_0 - v)^4}{Mv(1 - e^{-hv/kT})}[45(\bar{\alpha}')^2 + 7(\gamma')^2] \tag{35}$$

The intensity observed in practice will of course be proportional to the total number of scattering molecules in the sample. For a species in solution, this implies proportionality to the volume concentration.

One further point must be mentioned when observed intensities of different Raman lines are to be compared. The intensity expressions given above [equations (34) and (35)] refer to a particular normal mode. Where two (or three) modes are degenerate, each may be shown to give the same intensity, and so, as degenerate modes have the same frequency, the observed intensity of the Raman line will be increased by the factor 2 (or 3) as compared with that for a nondegenerate mode.

The results deduced above show that the observed intensity depends upon both the invariants $\bar{\alpha}'$ and $\gamma'$ of the derived tensor. If the degree of depolarization of the line in question is known, it provides a relation between $\bar{\alpha}'$ and $\gamma'$ which may be used to eliminate one or the other of them from the intensity expression. For all except totally symmetric modes, $\rho_n$ is necessarily $\frac{6}{7}$ for symmetry reasons.* In the case of totally symmetric modes, $\rho_n$ must be determined experimentally, except for molecules with spherical symmetry for which $\rho_n$ is necessarily zero. Equation (29) may be rewritten in the following form:

$$(\gamma')^2 = \frac{45(\bar{\alpha}')^2 \rho_n}{1 - 7\rho_n} \tag{36}$$

Using this to eliminate $(\gamma')^2$ from the intensity equations (34) and (35), we obtain

$$(I_n)_{obs} = KI_0 \frac{(v_0 - v)^4}{Mv(1 - e^{-hv/kT})}\left[45\frac{6(1 - \rho_n)}{6 - 7\rho_n}(\bar{\alpha}')^2\right] \tag{37}$$

and

$$(I_p)_{obs} = KI_0 \frac{(v_0 - v)^4}{Mv(1 - e^{-hv/kT})}\left[45\frac{6}{6 - 7\rho_n}(\bar{\alpha}')^2\right] \tag{38}$$

Thus, the observed intensity may be used to obtain the value of $(\bar{\alpha}')^2$.

* See the subsection, "Rules of Polarization," pp. 26–27.

## EXPERIMENTAL MEASUREMENT OF INTENSITIES

Certain factors that influence intensities as measured experimentally will be pointed out, but not discussed in detail, in this section. In the first place, the theoretical intensity expressions derived in the preceding section apply strictly to scattering at right angles to the direction of the incident light. In practice it is no easy matter to define the scattering angle exactly, and therefore so-called *convergence effects* arise. With appropriate experimental arrangements, these may be reduced so as not seriously to affect the significance of the measurements.

Secondly, the theoretical expressions apply to free molecules. Ideally, therefore, intensity determinations should be made upon gaseous samples at low pressures; however, this, even if not ruled out for other reasons, would mean that the intensities to be measured would be very low. In practice, therefore, most work is done with liquid samples. This introduces a number of complications.

The first is that *geometrical–optical effects* (reflections and refractions) will occur both at the entry of the exciting light into the sample and at the emergence of the scattered light from it. These effects will be different for samples of different refractive index and so will complicate the comparison of observed intensities. The second complication is due to so-called *internal effects*, arising from the fact that both the intrinsic scattering power of a molecule in a liquid and also the effective electric field strength of the exciting light at the position of the molecule will be affected by the surrounding molecules in its immediate vicinity. It is possible to make allowances for geometrical–optical effects,[16] but our present knowledge does not allow a reliable assessment of internal effects when comparing measured intensities for liquids.

## BOND-POLARIZABILITY THEORY OF INTENSITIES

We have seen that, according to Placzek's theory of vibrational fundamentals, the Raman intensity for a given mode is governed by the derived tensor $\alpha'$ of the molecule as a whole, just as the Rayleigh intensity is governed by the molecular polarizability tensor $\alpha$. The theory first proposed by Wolkenstein[17] and later given general mathematical formulation by Long[18] seeks to express $\alpha$ for the molecule as made up of polarizability contributions located in the various bonds of the molecule and $\alpha'$ as made up of contributions resulting from the stretchings and changes of orientation of the bonds appropriate to the particular normal vibrational mode in question. (A certain analogy with force fields may be here discerned, where the potential energy of the molecule is expressed as the sum of contributions corresponding to bond stretchings and changes of bond angles.)

The simple Wolkenstein theory involves the following two assumptions: (1) the *bond polarizability* of any particular bond is affected only by the stretching of that bond and is independent of the stretching of other bonds, and (2) bond polarizabilities are unaffected by changes of bond orientation. The second of these simplifying assumptions has been relaxed in a subsequent modification of the theory.[19]

It becomes necessary to introduce for each bond not only the two invariants $\bar{\alpha}'_{bond}$ and $\gamma'_{bond}$ of the *derived bond polarizability* with respect to bond stretching, but also the anisotropy invariant $\gamma_{bond}$ of the bond tensor itself. The mathematical treatment, which will not be presented here, then permits the calculation of the relative intensities of different vibrational modes from these bond parameters, *provided that the true nature of the normal coordinates (in terms of bond stretchings and changes of bond orientations) is known.*

In practice, of course, we start with the measured intensities and seek to calculate from them the polarizability parameters of the various bonds. In general, we are faced at once with the impossible situation that the number of unknown parameters exceeds the number of intensity data, but by choosing specially simple cases where only a small number of bond parameters is involved, this difficulty may be avoided or minimized. Nevertheless the situation in general still remains unsatisfactory because we do not have true normal coordinates, but only ones calculated from an assumed force field which (as we have seen previously) can be only an approximation to the truth.

In ignorance of the true force field and, hence, the true normal coordinates, it is not possible to test fully the validity of the Wolkenstein theory. Some measure of progress may, however, be made as follows. Instead of assuming a simplified force field and using it to calculate the intensities in terms of the bond-polarizability parameters, we may reverse the procedure and use the observed intensity data (along with the frequency data) to calculate a force field. The calculation must of course involve the assumption that the Wolkenstein theory is valid. In suitably chosen cases,[20] the total combined intensity and frequency data are sufficient to determine all the force-constant parameters of a *complete* quadratic field. If we find now that the calculation fails to give a real field, then we may be sure that the assumption of the validity of the Wolkenstein theory is a false assumption. If, on the other hand, a real field is obtained, then all we can conclude is that the Wolkenstein theory *may* be correct, though we have not proved that it is. Unless we have some further independent observed data which depend upon the force field and so can be used to test it, we can proceed no further.

The evidence so far available seems to indicate that, although the Wolkenstein theory of bond polarizabilities is not perfect, it represents a fairly good approximation in certain cases. Bearing this in mind, it may be used to obtain from observed intensities values of the polarizability parameters (e.g., $\bar{\alpha}'_{bond}$ with respect to stretching) of various bonds.

## RELATION OF BOND POLARIZABILITY TO BOND CHARACTER

When a bond is stretched, the polarizability associated with it changes. As the polarizability is concerned with electrons, it is reasonable to suppose that the magnitude of the change due to stretching is related to the extent to which electrons are involved in the bond, i.e., to degree of covalent character. Thus, in the case of a purely ionic bond with no bonding electrons, the polarizability will be the sum of the polarizabilities of the two atoms involved and will be (to a good approximation) independent of their distance apart. (In a higher approximation, account must be taken of the mutual polarization of each atom by the other, and this does give a slight dependence upon the separation distance.[21]) Accordingly, we shall expect an almost zero Raman intensity for the stretching of a purely ionic bond. For a covalent bond, on the contrary, the state of the bonding electrons (and, hence, the polarizability associated with them) will be strongly dependent upon changes of bond length, and so a correspondingly large Raman intensity will be anticipated.

The evidence in general bears out these expectations; Raman intensity measurements, interpreted on the basis of the Wolkenstein theory, have been used to obtain information about bond character regarding both degree of covalency and bond order. This is the theme of Chapter 4 of this book, which, however, sounds the warning that "great care must be taken in interpretation of bond-polarizability derivatives in terms of the distribution of electrons in a chemical bond."

## RAMAN SPECTRA OF SOLIDS

In the foregoing discussion, attention has been concentrated principally upon gases and liquids, in which the individual molecules can assume all orientations relative to the direction of the incident light. In a single crystal, the constituent molecules have fixed orientations, and there is no question of the sort of averaging undertaken in the case of fluids. It follows that the observed Raman spectrum will be dependent upon the overall orientation of the single-crystal sample relative to the direction of irradiation; in principle, therefore, it becomes possible to obtain information about the arrangement of the molecules in the crystal lattice.[22]

It must be borne in mind, however, that the spectra of molecules in crystals may be modified in two ways as compared with free molecules. First, if the unit cell of the structure contains more than one molecule, new modes of vibration due to coupling will make their appearance. Secondly, the effective symmetry of a single constituent molecule in the crystal may be lowered on account of its environment, and in consequence degeneracies may be removed and the selection rules of the free molecule may no longer be obeyed. For the spectroscopist who is a chemist interested in the structure of the free molecule, these features represent undesirable complications; on the other hand, for the investigator interested in crystal structure, they provide valuable new information. (For a general discussion of the vibrations of crystals, see the work of Mathieu.[23])

A crystal powder, in which the particles are randomly oriented, is in this respect more akin to a liquid sample, although the remarks of the preceding paragraph still apply. From an experimental point of view, however, the inevitable reflection of intense primary light by the powder makes it more difficult to obtain Raman spectra of high quality, especially for the smaller frequency shifts. Nevertheless, by using a spectrograph with an efficient discrimination between exciting and Raman-scattered light, useful spectra can be obtained for substances which, for one reason or another, cannot be studied in the liquid state or in solution.

## SPECIAL EFFECTS WITH EXTREMELY HIGH-INTENSITY LASER EXCITATION

All that has been written in the foregoing sections refers to light-scattering when the incident intensity is in the normal range obtainable either from conventional sources or from lasers. The Raman scattering is then essentially incoherent in character.

By suitable regulation (known as $Q$-switching), the emission from a ruby laser can be obtained in a single "giant pulse" of very short duration (of the order of $10^{-8}$ sec) and of very high peak power (up to 100 MW or more). When such an extremely intense flash of coherent light is incident upon a sample, entirely novel phenomena are observed.[24–26] The quantum-mechanical theory of the normal Raman effect becomes inadequate, as does also the polarizability theory in the form in which it has been presented earlier in this chapter.

### Stimulated Raman Effect

With giant-pulse excitation of frequency $v_0$, a gain is induced in the sample at a certain Stokes frequency $v_0 - v$, where $v$ is the frequency of a Raman-active vibration. Usually only one such frequency is "active" in this way, namely, the one having the largest normal Raman intensity per line width; however, occasionally two vibrations may take part simultaneously. Such vibrations always belong to totally symmetric species, since this type is characterized by the most intense and narrowest normal Raman lines. Provided the incident intensity exceeds a certain *threshold*, the gain exceeds the losses, and a stimulated Stokes line is emitted with an intensity (up to 20% of that of the exciting laser beam) which is very high indeed as compared with normal Raman scattering. This stimulated emission, unlike the normal Raman effect, is coherent.

At the same time, other lines appear on the Stokes side of $v_0$. The frequency shifts therefrom are exactly twice, thrice, four times, *etc.*, that of the first line, and their intensities are lower and fall off progressively. They are referred to as higher-order stimulated Stokes lines. Under suitable experimental conditions, corresponding lines are obtained on the anti-Stokes side. Their intensities are lower than those for the Stokes analogs, but still very much higher than those for anti-Stokes lines in the normal Raman effect. Experiment shows that the emission of the higher-order Stokes lines and the anti-Stokes lines is specifically directional.

These phenomena, referred to generally as the *stimulated Raman effect*, are not yet fully understood. According to the classical theory advanced by Garmire, Panderese, and Townes,[27] which is supported by the observations of Stoicheff,[25] diffuse first-order Stokes radiation (frequency, $v_0 - v$) is initially produced, and it is the interaction of this with the incident radiation which gives rise to the higher-order Stokes and the anti-Stokes lines.

Since in most cases only one vibrational frequency appears (together with its multiples) in the stimulated Raman spectrum, the effect is clearly of little value to the molecular spectroscopist who wishes to determine as many fundamental frequencies as he can in order to characterize the scattering molecule or to obtain data for force-field calculations. The principal practical importance of the effect appears rather to lie in the method it provides for obtaining, from an original laser beam, intense coherent light of shifted frequency.

### Inverse Raman Effect

The quantum theory of the normal Raman effect predicts, in addition to emission, the possibility of absorption at the shifted frequencies. The existence of such absorption was demonstrated experimentally for the first time in 1964 by Jones and Stoicheff.[28] A sample S (benzene) was irradiated with laser light of frequency $v_0$ and of intensity below the threshold for the stimulated Raman effect. At the same time it was irradiated by a very intense continuum covering the position of an anti-Stokes Raman line. The continuum used was in fact a stimulated anti-Stokes feature from a different substance S' (toluene), chosen so as to cover the desired frequency region when produced under conditions giving very considerable line broadening. By this device a sharp absorption was observed at the selected anti-Stokes frequency of S. The mechanism of its production is believed to be related to that of the stimulated Raman effect. No threshold is expected, however, and so in principle the complete Raman spectrum (as normally obtained in emission) should be observable in absorption. In practice, this must depend upon the possibility of realizing the requisite extended continuum of very high intensity. At present there appears to be no way of producing this.

The potential importance of the inverse Raman effect is great, especially in view of the extremely short time (one pulse of the continuum) required to observe it. If the experimental difficulties can be overcome, it may furnish a new type of high-speed Raman spectroscopy applicable, for example, to the study of short-lived species, such as free radicals.

## REFERENCES

1. C. V. Raman and K. S. Krishnan, *Nature* 121: 501 (1928); *id.*, *Proc. Roy. Soc. (London)* 122A: 23 (1928). C. V. Raman, *Indian J. Phys.* 2: 387 (1928); C. V. Raman and K. S. Krishnan, *ibid.*, p. 399.
2. A. Smekal, *Naturwissenschaften* 11: 873 (1923).
3. G. Landsberg and L. Mandelstam, *Naturwissenschaften* 16: 557 and 772 (1928).
4. H. Stammreich, *Spectrochim. Acta* 8: 41 (1956–57).
5. J. R. Nielsen, *J. Opt. Soc. Am.* 20: 701 (1930); *id.*, *ibid.*, 37: 494 (1947).
6. D. H. Rank, R. J. Pfister, and P. D. Coleman, *J. Opt. Soc. Am.* 32: 390 (1942); D. H. Rank, R. J. Pfister, and H. H. Grimm, *ibid.*, 33: 31 (1943); D. H. Rank and R. V. Wiegand, *ibid.*, 36: 325 (1946).
7. G. Placzek, *Rayleigh-Streuung und Raman-Effekt,*" in: E. Marx, *Handbuch der Radiologie*, Akademische Verlagsgesellschaft, Leipzig, 1934, Vol. 6, Part 2, p. 205.
8. M. S. Child and H. C. Longuet-Higgins, *Phil. Trans. Roy. Soc. London Ser. A* 254: 259 (1961).
9. G. Placzek and E. Teller, *Z. Physik* 81: 209 (1933).
10. B. P. Stoicheff, "High-Resolution Raman Spectroscopy," in: H. W. Thompson, *Advances in Spectroscopy*, Interscience, New York, 1959, Vol. 1, p. 91.
11. E. P. Wigner, *Group Theory*, Academic Press, New York, 1959.
12. E. B. Wilson, J. C. Decius, and P. C. Cross, *Molecular Vibrations*, McGraw-Hill, London, 1955, p. 323 *et seq.*
13. G. Herzberg, *Infra-red and Raman Spectra*, Van Nostrand, New York, 1945 (reprinted 1947).
14. L. Pauling and E. B. Wilson, *Introduction to Quantum Mechanics*, McGraw-Hill, New York, 1935, p. 82.
15. E. B. Wilson, J. C. Decius, and P. C. Cross, *Molecular Vibrations*, McGraw-Hill, London, 1955.
16. D. G. Rea, *J. Opt. Soc. Am.* 49: 90 (1959).
17. M. W. Wolkenstein, *Compt. Rend. Acad. Sci. U.R.S.S.* 32: 185 (1941); M. Eliashevich and M. W. Wolkenstein, *J. Phys. (Moscow)* 9: 101 and 326 (1945).
18. D. A. Long, *Proc. Roy. Soc. (London)* 217A: 203 (1953).
19. D. A. Long, A. H. S. Matterson, and L. A. Woodward, *Proc. Roy. Soc. (London)* 224A: 33 (1954).
20. G. W. Chantry and L. A. Woodward, *Trans. Faraday Soc.* 56: 1110 (1960); K. A. Taylor and L. A. Woodward, *Proc. Roy. Soc. (London)* 264A: 558 (1961).
21. J. H. B. George, J. A. Rolfe, and L. A. Woodward, *Trans. Faraday Soc.* 49: 375 (1953).
22. L. Couture-Mathieu and J. P. Mathieu, "The Orientation of Water Molecules in a Crystal of a Hydrated Salt," *Acta Cryst.* 5: 571 (1952).
23. J. P. Mathieu, *Spectres de Vibration*, Hermann, Paris, 1945, Chapters 11 and 12.
24. G. Eckhardt, R. W. Hellwarth, F. J. McClung, S. E. Schwarz, and D. Weiner, *Phys. Rev. Letters* 9: 455 (1962).
25. B. P. Stoicheff, *Phys. Letters* 7: 186 (1963).
26. R. Chiao and B. P. Stoicheff, *Phys. Rev. Letters* 12: 290 (1964).
27. E. Garmire, F. Panderese, and C. H. Townes, *Phys. Rev. Letters* 11: 160 (1963).
28. W. J. Jones and B. P. Stoicheff, *Phys. Rev. Letters* 13: 657 (1964).

*Chapter 2*

# Advances in Raman Instrumentation and Sampling Techniques

## John R. Ferraro

*Argonne National Laboratory*
*Argonne, Illinois*

### INTRODUCTION

In 1928, when Sir Chandrasekhara Raman discovered the phenomenon which bears his name, the instrumentation was very crude. Raman used sunlight as the source, a telescope as the collector, and his eyes were the detector. It was very remarkable that such a small phenomenon as the Raman effect was detected with such primitive instrumentation.

Gradually, there was improvement in the various components of Raman instrumentation. Early research centered on the development of better sources. Research workers developed cadmium, helium, thallium, bismuth, lead, and zinc lamps.[1-3] However, these proved to be unsuitable because of low light intensities. Attention then turned to the mercury lamp. A water-cooled mercury lamp was first introduced in 1914 by Kerschbaum.[4] Mercury lamps applicable to Raman use were developed in the thirties.[5] Hibben developed a mercury burner in 1939,[6] and Spedding and Stamm[7] experimented with a cooled mercury burner in 1942. Further progress was made by Rank and McCartney[8] in 1948 when they studied mercury burners and their background conditions. A commercial mercury lamp was developed by Hilger for their Raman instrument. The source consisted of four lamps, which surrounded the Raman tube. A most important development occurred in 1952 when Welsh *et al.*[9] introduced a mercury source, which since has been called the Toronto arc. The lamp consists of a four-turn helix of Pyrex tubing and is more intense than the Hilger unit. Further improvements in lamps have occurred recently. Ham and Walsh[10] described the use of helium, mercury, sodium, rubidium, and potassium lamps. These are the annular electrodeless lamps powered by microwaves at 2450 Mc. The search for more intense Raman sources has continued, and within the last few years laser sources applicable to Raman spectroscopy have made their appearance. These new laser sources will be discussed in a later section of this paper. Improvements also were made in the filters in order to provide a relatively monochromatic source. A good review of these is provided by Brandmüller and Moser.[11]

Eventually, several workers in Raman spectroscopy constructed homemade instruments. The early instruments recorded photographically and were time-consuming and cumbersome. In addition, few of these instruments provided satisfactory discrimination of the Rayleigh scattering of the exciting line. Since this paper attempts to discuss recent instrumental developments, only photoelectric Raman instruments will be discussed. In the following discussion several of these instruments will be highlighted.

## RAMAN INSTRUMENTATION

### The Advent of Photoelectric Raman Spectrophotometers

The first photoelectric Raman instruments began to appear following World War II. The first of these was reported in 1946 by Rank and Wiegand.[12]

**The Rank and Wiegand Instrument.** In 1942, Rank and co-workers[13] demonstrated that Raman spectra could be obtained photoelectrically by using an RCA 931 multiplier phototube as a detector. In 1946, Rank and Wiegand[12] introduced the first photoelectric Raman spectrophotometer. The sources used were two Westinghouse H-1 mercury vapor lamps (400 W), and the filter was an aqueous $NaNO_2$ solution. The detector was an RCA IP21 phototube.

The spectrograph has a concave grating on a Wadsworth mounting. The grating is ruled at 15,000 lines/in. The light from the entrance slit is collimated by means of a 10-in.-aperture, 94.5-in.-focal-length parabolic telescope mirror, which is front-surface aluminized. The parallel light from the collimator mirror falls on an 8-in.-diameter, 15-ft-radius of curvature concave grating. The light is diffracted into the second-order spectrum in the region of 4000–5000 Å and brought to focus on the exit slit. The spectrum is scanned by rotation of the diffraction grating. The cascade type RCA IP21 detector is mounted in a can which is cooled to dry-ice temperature. Light from the exit slit is projected through the double windows of the photocell housing by means of a single equiconvex lens. The photocell, photocell refrigerant, lens, and exit slit are mounted on a platform floated on ball bearings. A photographic galvanometer deflection recorder is used.

**The Heigl Instrument.** In 1950, another photoelectric Raman spectrophotometer was introduced by Heigl and co-workers.[14] The Heigl instrument uses a Toronto-type low-pressure arc, with a Young–Thollen prism monochromator and a refrigerated RCA-C-7073B photomultiplier. The instrument scans 3500–3550 cm$^{-1}$ from the 4358-Å exciting line. It has an aperture of $f/8.8$ and a dispersion of 23 Å/mm at 4861 Å. The prism is of DF-2 glass, which has high dispersion in the violet region. The prism assembly consists of one 60° and two 30° prisms. The 60° prism is mounted stationary, while the two 30° prisms are kinematically mounted and can be rotated by a motor-driven wedge. The instrument uses aqueous $NaNO_2$ solution as the filters and a Brown-Electronik recorder.

**Raman Instruments Designed by Stamm and Co-Workers.** In 1953, Stamm and co-workers[15] reported the conversion of a photographic Raman instrument into a photoelectric Raman spectrophotometer. The instrument used a Toronto-type mercury arc, with a filter jacket containing aqueous $NaNO_3$ solution and rhodamine. The spectrograph is similar to the one previously discussed[16] and has a Gaertner bilateral slit, an off-axis paraboloidal collimating mirror, a plane echelette grating ruled at 15,000 lines/in., and a fast objective lens. The instrument features a scanning-slit mechanism. The detector is an RCA IP21 photomultiplier tube, refrigerated with liquid nitrogen. For the refrigeration, a vacuum chamber is necessary and a stainless Dewar with a 3-gal capacity is used. A conventional power supply is used, which delivers an output voltage of 600–1500 V (actually 490 V has been found to achieve a suitable regulation).

In 1953, Stamm and Salzman[17] converted a Perkin-Elmer Model 12 spectrophotometer to a photoelectric Raman instrument. The conversion was made to a grating-type spectrophotometer, with changes made in the slits, mounting for a paraboloid and a wavelength drive. The grating is a plane echelette type ruled 15,000 lines/in. The instrument uses a Toronto-type mercury four-turn helix lamp source, made of 25-mm OD Pyrex. The lamp is operated at 15 A

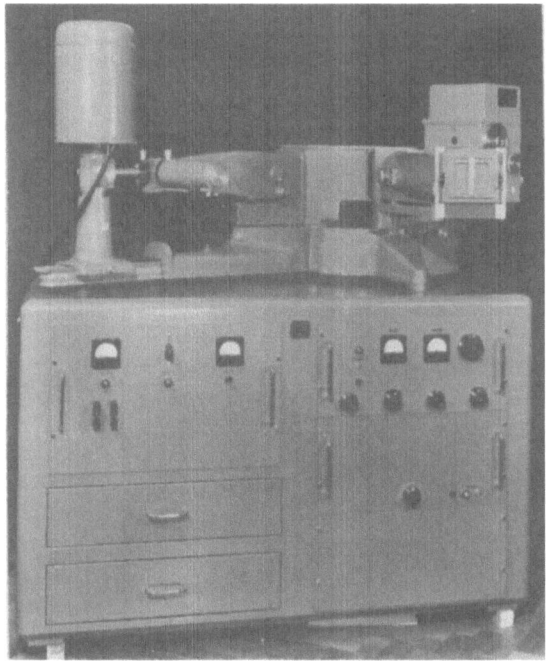

Fig. 1. Hilger E612 Raman spectrometer.

and 95 V DC and is backed with a chromium plate reflector. The instrument contains a filter jacket utilizing aqueous $NaNO_2$ solution. The instrument features image slicers[18–20] made of multiple-prism assemblies. The power supply and detector are the same as previously discussed for the other Stamm instrument.

**The Hilger Instrument (E612).** The Hilger E612 (see Figs. 1 and 2) can be operated as either a photographic or photoelectric instrument. The operation of the instrument as a photoelectric Raman instrument will be described. The instrument is a two-prism spectrograph. It has a collimator, with an added separate curved exit slit and glass lens. Two glass prisms are mounted in a rigid metal base. The collimator has a lens aperture of 86 mm and a focal length of 26 in. The glass prisms have refracting angles of 63° and are 86 mm high and 130 mm wide, with a 164-mm face length. An EMI photomultiplier tube is used as a detector. The source unit consists of four mercury vapor lamps. The light is dispersed in the first pass through the prisms and reflected at a filtering mirror to obtain double dispersions

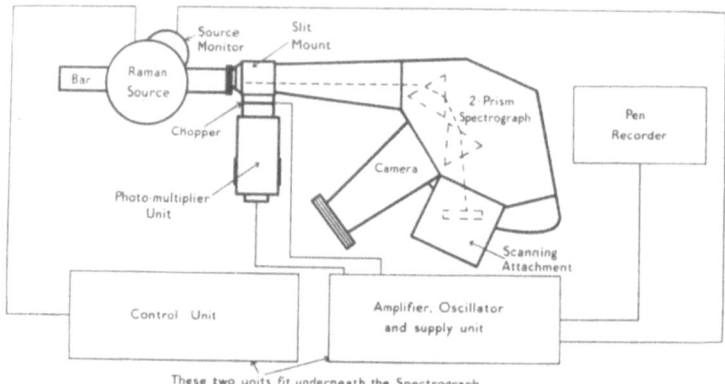

Fig. 2. Hilger E612 optical diagram.

(6.8 cm$^{-1}$/mm at 4358 Å). The mirror is then rotated automatically at one of four selected speeds. The signal goes through an exit slit, is interrupted by a chopper, and is then brought to focus on the photo-multiplier. The signal for the photomultiplier is then amplified and operates a pen recorder.

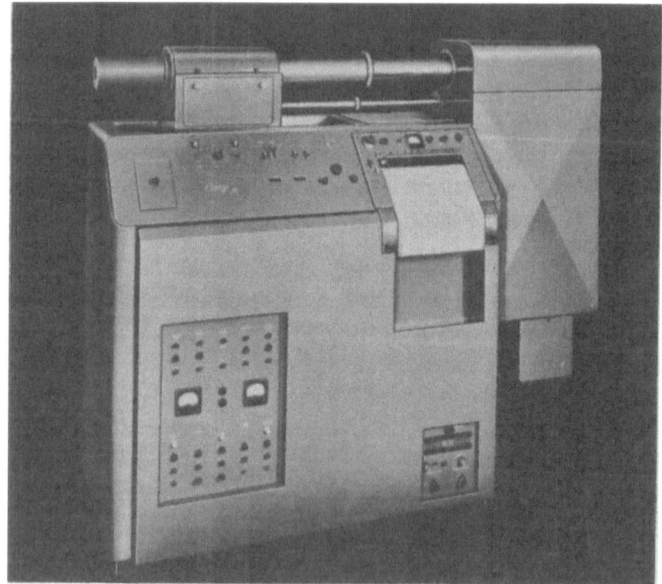

Fig. 3. The Cary Model 81 Raman spectrophotometer.

**The Cary Model 81 Raman Spectrophotometer.** The instrument which gave Raman spectroscopy a tremendous boost in the United States was the Cary Model 81 Raman spectrophotometer. The instrument was first described at the Ohio State University Symposium,

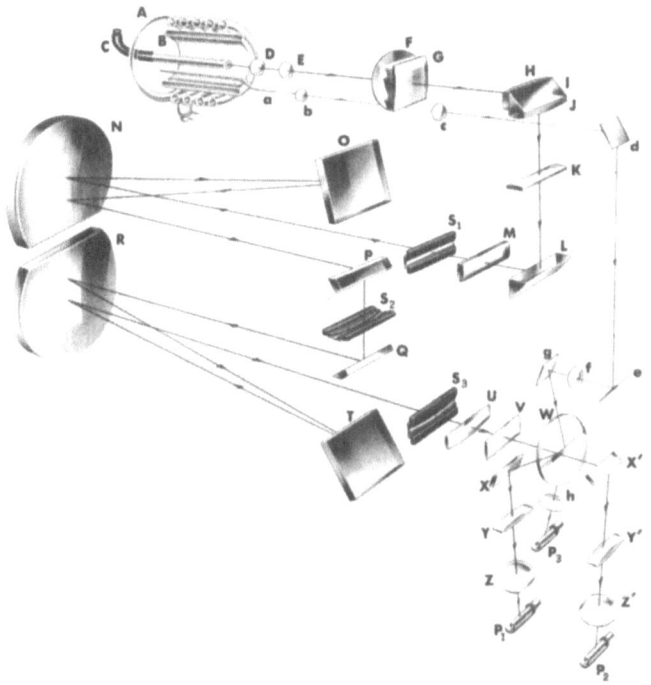

Fig. 4. The Cary Model 81 optical diagram.

Light from Toronto-type arc lamp A is filtered by fluid in jacket B so that only energy of 4358-Å wavelength illuminates the sample contained in cell C. Raman light from the end of the sample cell is directed by lenses D and E through collimating lens F to the first image slicer G, which divides the beam into 20 images. These are directed into the 10 sections of the second image slicer H, each section receiving two images. The second image slicer, with the aid of lenses J, K, and M, superimposes images of the elements of G into two narrow strips of light in the plane of double entrance slit $S_1$. Prisms I and L serve only to change the direction of the light. From collimating mirror N, the beam is reflected to grating O and then reflected by mirror P through double intermediate slit $S_2$ to the second monochromator. From $S_2$ it is reflected by mirror Q and the second collimator R to the second grating T, and then directed through double exit slit $S_3$. Lenses U and V direct the beam to rotating mirror W which alternately directs it to phototubes $P_1$ and $P_2$ by means of lenses X, Y, Z and X', Y', Z'. The two phototube signals are combined and compared to the reference signal developed by phototube $P_3$. Concurrently, part of the filtered lamp radiation is directed by glass light pipe a into an auxiliary optical train (elements b, c, d, e, f, g) to rotating mirror W. It is then directed through lens h to reference phototube $P_3$ which develops a signal for comparison to the Raman signals from phototubes $P_1$ and $P_2$.

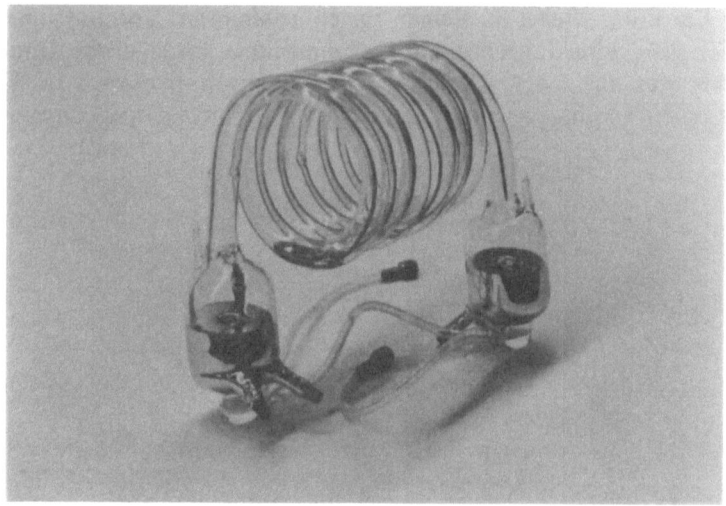

Fig. 5. The mercury arc lamp used in the Cary Model 81.

June, 1953.[21] The relative ease of operation, sensitivity, high reproducibility, and high resolution are features on this instrument.

The instrument and its optical diagram are shown in Figs. 3 and 4, respectively. The source used in the Cary 81 is a 3-kW helical mercury arc of the Toronto type and operates on 18 A of current and a voltage of 150 V. The lamp source (Fig. 5) is made in the form of a helix from Corning 1720 glass tube and it surrounds the sample cell. The lamp is cooled by air from two blowers, and the electrodes are cooled with circulating thermostatted water.

A cylindrical glass filter jacket is placed between the lamp and the sample cell. The filter solution circulates through the jacket and transmits the 4358-Å mercury line while absorbing other interfering mercury lines. The filter solution is cooled, and this in turn cools the sample. A description of the Cary 81 sample cells and their use will be given in a later section.

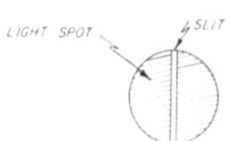

Fig. 6. Cross-sectional diagram of the beam of Raman radiation falling on the entrance slit to the monochromator in the Cary Model 81. With circular apertures, a circular spot of Raman energy falls on the entrance slit, and only that which can pass through the slit enters. Much of the energy is thus lost (cross-hatched area). The image slicers prevent this waste of Raman energy.

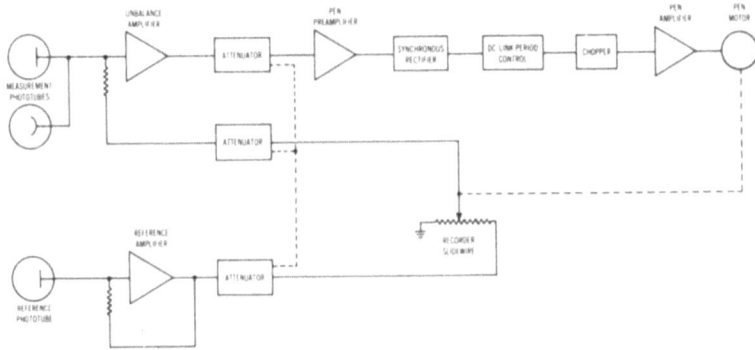

Fig. 7. Model 81 circuit block diagram.

The use of image slicers increases the sample light efficiency. The image slicers make it possible to use the light from the entire end of the sample cell, instead of only a narrow strip in the center. This is accomplished by slicing the image at the end of the cell into 20 strips, then superimposing them into two sets of 10 strips. The two images are then magnified to the size that fills the double entrance slit. Normally, in a conventional optical train, only a very small portion of the energy enters through the slit into the monochromator, illustrated by Fig. 6. Thus, the image slicers make use of a large portion of

## TABLE I

### Comparison of Figures of Merit for Various Photoelectric Raman Spectrophotometers

| Instrument | $T^a$ | $A^b$ (cm²) | $D^c$ (× 10⁴) (rad/Å) | $H^d$ (cm) | $F^e$ (cm) | $H/F$ | $Q_1^f$ (× 10⁵) | $Q_2^g$ (× 10⁴) | $Q_3^h$ (× 10⁴) |
|---|---|---|---|---|---|---|---|---|---|
| ARL | 0.24 | 49 | 1.6 | 1.6 | 92 | 0.017 | 3.3 | 3.0 | 0.95 |
| Heigl | 0.65 | 28 | 0.65 | 2.2 | 66 | 0.033 | 4.0 | 2.9 | 0.94 |
| Rank | 0.36 | 147 | 1.26 | 1.0 | 240 | 0.004 | 2.8 | 3.0 | 1.23 |
| Stamm I | 0.44 | 137 | 0.63 | 0.6 | 100 | 0.006 | 2.3 | 2.7 | 0.86 |
| Stamm II | 0.32 | 56 | 1.3 | 0.71 | 27 | 0.026 | 4.0 | 2.9 | 0.94 |
| APC (Cary 81) | 0.18 | 100 | 2.5 | 20 | 100 | 0.2 | 91 | 34 | 6.4 |

[a] $T$ is a transmission factor.
[b] $A$ is the area of the stop limiting the aperture of the prism or grating.
[c] $D$ is the angular dispersion of the monochromator.
[d] $H$ is the entrance slit height.
[e] $F$ is the entrance focal length.
[f] $Q_1 = TADH/F$.
[g] $Q_2 = T(ADH/F)^{\frac{1}{2}}$.
[h] $Q_3 = TA^{\frac{1}{2}}DH^{\frac{1}{2}}/F^{\frac{1}{2}}$.

the Raman scattering that otherwise would never enter the entrance slit.

The monochromator is a twin-grating, twin-slit, double monochromator. The gratings used are ruled at 1200 lines/mm and blazed for high efficiency in the first order at 4500 Å. They are collimated by spherical mirrors operated off-axis in a Czerny–Littrow arrangement. The double monochromator permits the use of a large slit height-to-focal length ratio and the Shurcliff double-slit system increases the light-gathering power of the instrument. This arrangement considerably aids in minimizing the Tyndall and Rayleigh scattering in the sample.

The Raman radiation from the exit slit of the monochromator is conducted to a rotating semicircular mirror, which causes the radiation to be directed alternately to two phototubes (IP-28) at a 30-cps frequency. The two phototubes produce signals of opposite polarity and are combined and applied to the input of the unbalance amplifier, as indicated in Fig. 7. Simultaneously, nondispersed light from the lamp is chopped by the semicircular mirror and directed to a reference phototube. This reference signal is amplified and directed to an attenuator and a recorder slidewire. The signal is then applied to the input of the unbalance amplifier. Here the signal is combined with that from the measurement phototubes, and if the two signals are equal, they cancel out. A difference in signals is amplified, rectified to DC, chopped, and used to drive a motor which operates a recorder pen.

Table I compares the figures of merit for various Raman photoelectric spectrometers. It can be observed from the table that the figures are rather uniform for the first five instruments. The table illustrates the outstanding features of the Cary Model 81, where $H/F$ is 0.2 and the values of $Q_1$, $Q_2$, and $Q_3$ are much higher than those of the other instruments.

## Laser Raman Spectrophotometers

The search for a more intense monochromatic source for the Raman effect has been the task of many scientists ever since Raman discovered the phenomenon. The application of the use of a laser for a Raman source has been researched from the moment the laser was invented. The first reported Raman source of this type was the pulsed ruby laser.[22] This source had disadvantages in that it required

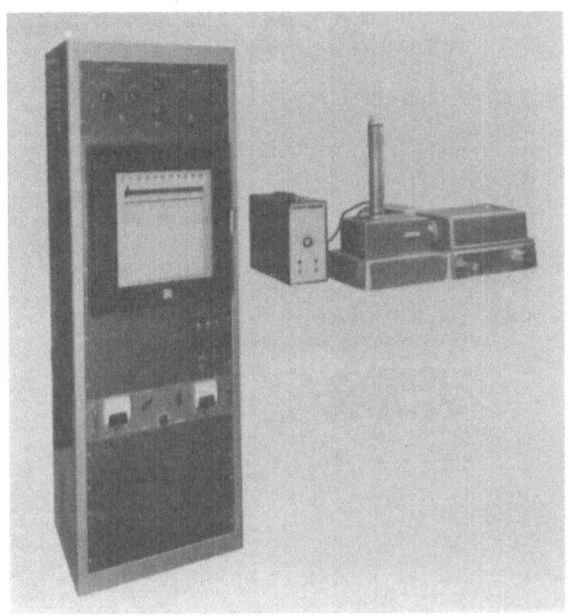

Fig. 8. Perkin-Elmer laser Raman spectrophotometer source
mounted on top of instrument.

Fig. 9. Close-up of Perkin-Elmer laser Raman spectrophotometer source
mounted on top of instrument.

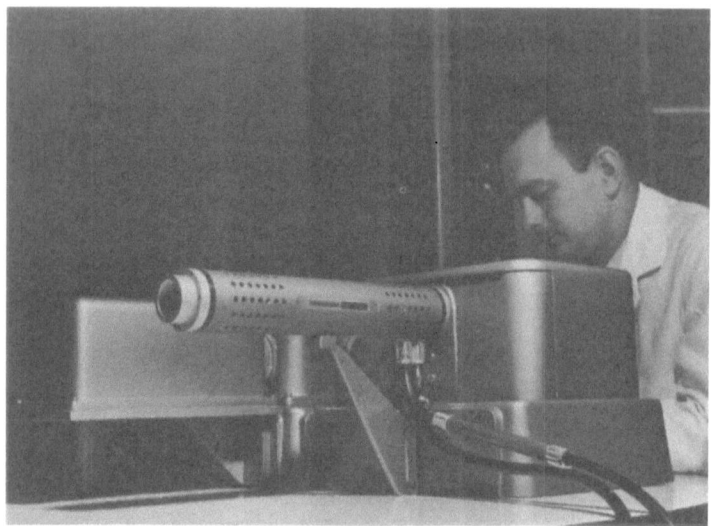

Fig. 10. Perkin-Elmer laser Raman spectrophotometer source mounted on side of instrument.

Fig. 11. Close-up of Perkin-Elmer laser Raman spectrophotometer source mounted on side of instrument.

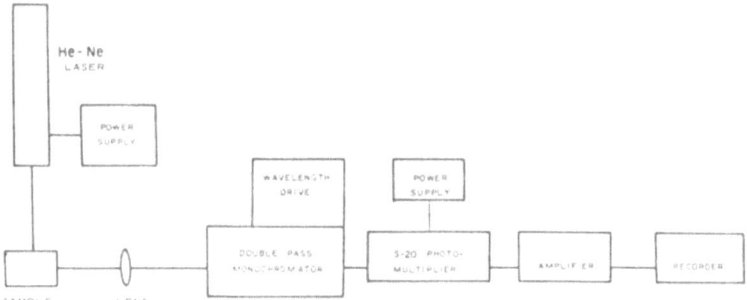

Fig. 12. Block diagram of laser Raman spectrophotometer.

50–100 laser pulses to give a reasonable Raman effect and had to be cooled with liquid nitrogen. In addition, a large amount of unwanted energy entered the spectrograph, and the instrument had poor Raman light-gathering power. The use of a helium–neon gas laser oscillating at 6328 Å as a Raman source was first reported in 1963 by Kogelnik and Porto.[23] In 1964, the method was adapted to utilize photoelectric recording techniques and a commercial instrument was developed by the Perkin-Elmer Corporation.[24]

**Perkin-Elmer Laser Raman Spectrometer.** Very recently, the Perkin-Elmer Corporation designed a laser-excited Raman spectrophotometer, which records photoelectrically. The instrument uses a helium–neon gas laser of about 4-mW power. The monochromator is a double pass optical system. A special end-viewing photomultiplier tube with an S-20 response photocathode serves as the detector. The other components consist of an amplifier, a standard wavelength

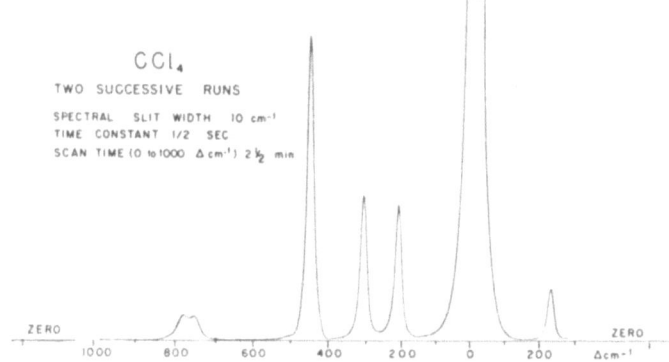

Fig. 13.  Laser Raman spectrum of $CCl_4$.

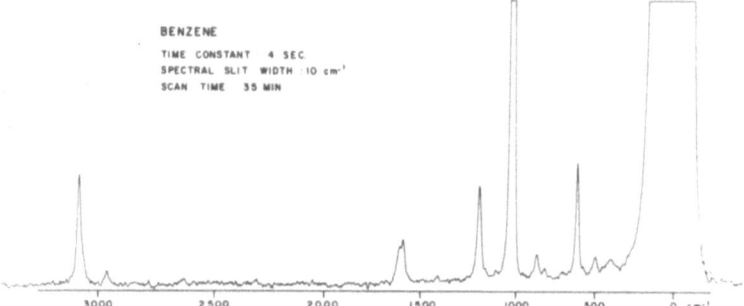

Fig. 14.  Laser Raman spectrum of benzene.

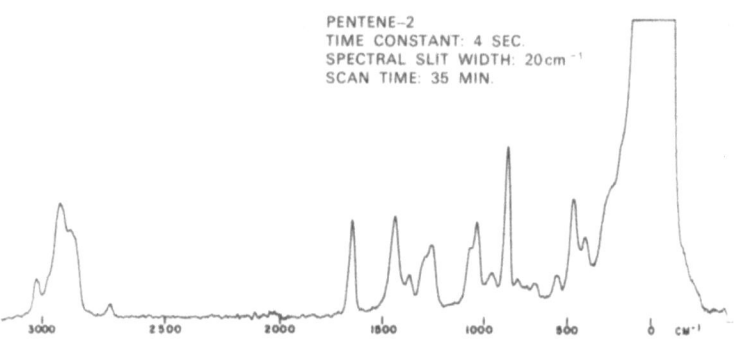

Fig. 15.  Laser Raman spectrum of pentene-2.

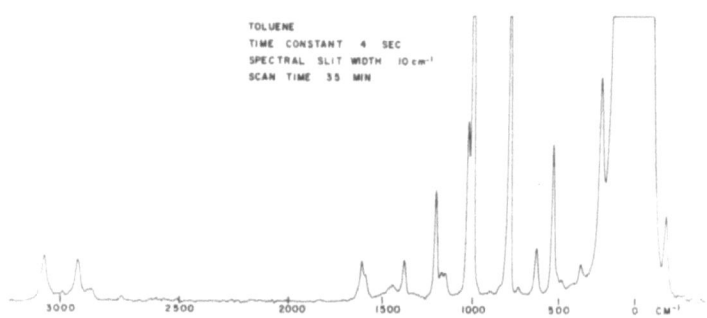

Fig. 16.  Laser Raman spectrum of toluene.

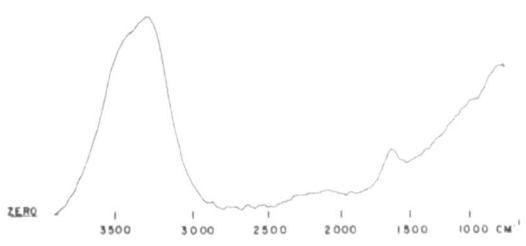

Fig. 17. Laser Raman spectrum of $H_2O$.

drive and control unit, and a Leeds & Northrup strip chart recorder. The entire assembly of the Perkin-Elmer laser Raman spectrophotometer, as first introduced, is shown in Figs. 8 and 9. Recently the position of the source was changed 90° in order to accommodate low-temperature Dewars in the sample area. Figures 10 and 11 illustrate this change. Figure 12 is a block diagram of the instrument.

The 6328-Å red laser emission line is used. The laser consists of a double-walled, gas discharge tube placed inside a coaxial cylinder housing. By means of a small DC discharge between an internal anode and heated cathode, plasma is generated inside of a capillary. The end of the tube has Brewster's angle windows, which causes only

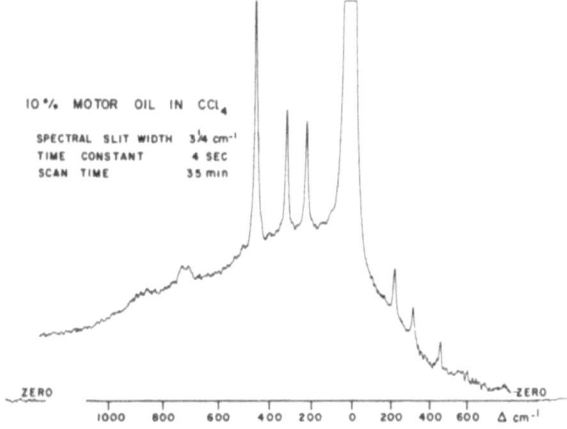

Fig. 18. Laser Raman spectrum of 10% motor oil in $CCl_4$.

## TABLE II

### Comparison of Various Photoelectrical Raman Spectrophotometers with Regard to Several Components

| Instrument | Source | Spectrograph | Detector | Recorder |
|---|---|---|---|---|
| Rank | 2-H-1 Hg vapor lamps | Grating monochromator 15,000 lines/in. | RCA IP21 (refrigerated with dry ice) | Photographic galvanometer deflection recorder |
| Heigl | Toronto-type low pressure | Three-prism | RCA-C-7073B (refrigerated) | Brown–Electronik |
| Stamm I | Toronto-type | Grating 15,000 lines/in. | I-P21 (refrigerated with liquid $N_2$) | Leeds & Northrup Speedomax G |
| Stamm II | Toronto-type | Grating 15,000 lines/in. | I-P21 (refrigerated with liquid $N_2$) | Leeds & Northrup Speedomax G |
| ARL | Toronto-type | Three-prism | I-P21 | Brown–Electronik |
| Hilger | Four low-pressure Hg lamps | Two-prism | EMI | Cambridge Recorder or Brown–Electronik |
| APC (Cary 81) | Toronto | Double prism and grating 1200 lines/mm | IP28 | APC |
| Perkin-Elmer | He–Ne laser | Double grating 1440 lines/mm | S-20 | Leeds & Northrup Speedomax G |
| APC laser Raman | He–Ne laser | Double grating blazed at 7500 Å | S-20 | APC |

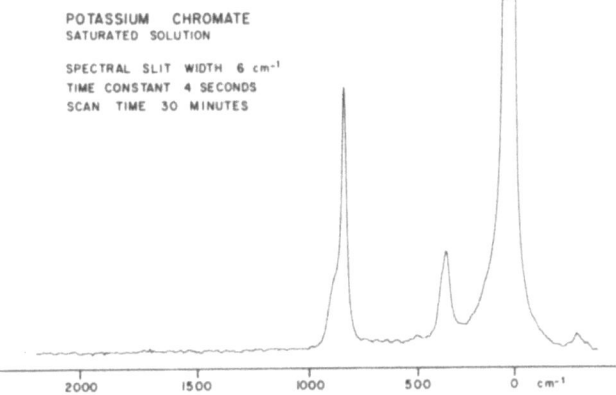

Fig. 19. Laser Raman spectrum of potassium chromate.

plane-polarized light to be emitted. External mirrors are coated with a dielectric material, which maximizes the 6328-Å line.

The laser beam goes through a specially designed sample cell, which will be discussed in a later section under sampling techniques. A small collecting spherical mirror, placed behind the cell, gathers the back-scattered Raman energy and allows it to focus toward the monochromator entrance slit. The monochromator has a double pass system and a 13 cps chopper. The grating is ruled at 1440 lines/mm. Scattered radiation other than Raman scattering that is centered about the same wavelength as the exciting line is not chopped

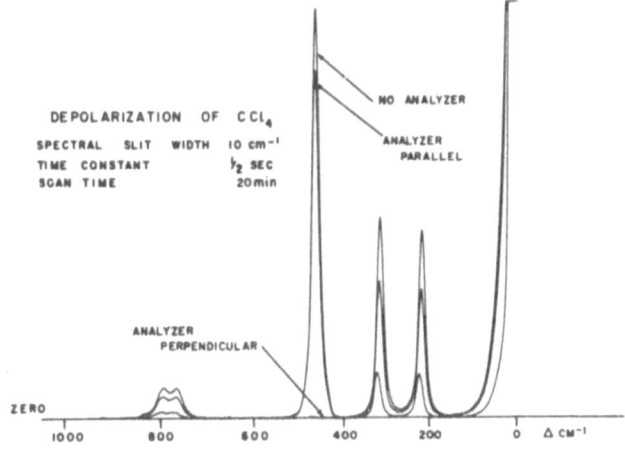

Fig. 20. Depolarization of $CCl_4$ with the laser Raman.

on the first pass through the monochromator and is, therefore, not seen by the detector.

The optics in the monochromator are all front-surface-coated with a 99% reflecting dielectric material in the region 6300–8300 Å. This improves the transmission through the system by four times that available from aluminum mirrors. To maximize the signal-to-noise ratio, an end-viewing photomultiplier tube with an S-20 response photocathode tube is used.

Figures 13–16 are typical spectra obtainable by the Perkin-Elmer laser Raman spectrophotometer for $CCl_4$, benzene, pentene-2, and toluene, respectively. Figure 17 is a spectrum of water. Figure 18 is the spectrum of $CCl_4$ containing 10% motor oil. Although fluorescence occurs, the spectrum of $CCl_4$ is easily observed. The spectrum of a saturated solution of potassium chromate, a yellow color, is shown in Fig. 19. Figure 20 illustrates the spectra of $CCl_4$ with and without analyzers.

Table II compares the various photoelectrical Raman spectro-photometers with regard to source, monochromator, detector, and recorder.

**Laser Raman Model 81.** At the Pittsburgh Analytical Conference in March, 1965, Dr. R. C. Hawes of Applied Physics Corporation discussed the use of laser excitation with the Cary 81. The laser was a helium–neon Spectra-Physics Model 112 gas laser, having about 35 mW output multianode or 15 mW single mode at 6328 Å. The laser is focused coaxially into a capillary cell. The scattered light is collected

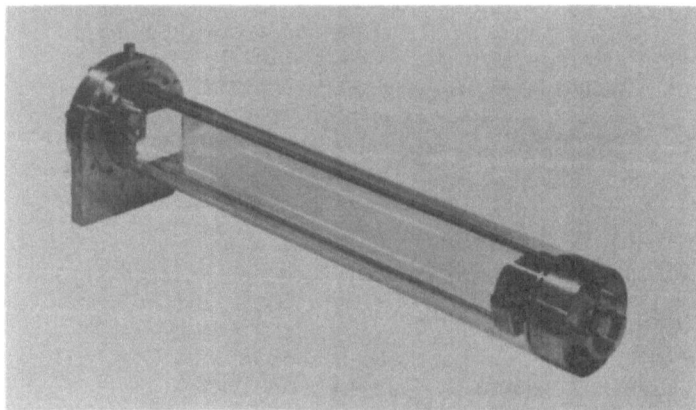

Fig. 21. Cary Model 81 multipass gas cell.

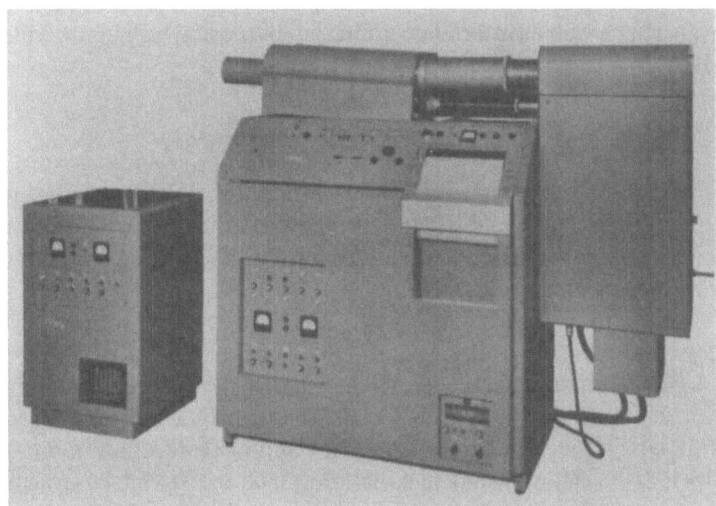

Fig. 22. Cary Model 81 with gas cell unit.

by the monochromator and detected by an S-20 photomultiplier tube. Larger gratings blazed at 7500 Å were used to improve efficiency. Improved coatings on both transmitting and reflecting optics were

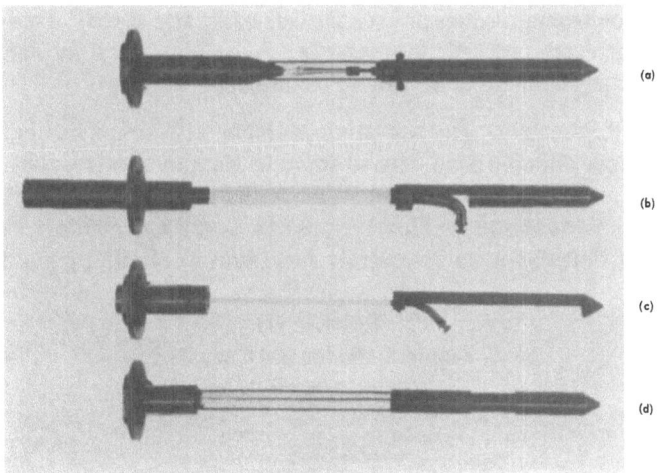

Fig. 23. Model 81 cells, holders, and cell optics. (a) Solid holder. (b) 19-mm liquid cell. (c) 7-mm liquid cell. (d) 2-mm liquid cell.

also used, as well as interference filtering of the exciting light to further reduce unwanted source radiation.

## SAMPLING TECHNIQUES

Most of the recent sampling techniques have been developed in relation to the Cary 81 and the new Perkin-Elmer laser Raman. As a result, most of the discussion that follows will be in terms of these new instruments.

### Cary Model 81

**Gas Sampling.** Gas cells such as the one shown in Fig. 21 are used to obtain Raman spectra of gases. This cell is a multiple-reflection cell in which the light can bounce back and forth 44 times in the 20-in. path length. The cell has a volume of 3500 cc. It can be evacuated, operated at atmospheric pressure, or pressurized to 10 atm. In order to utilize the gas cell, a special compartment is needed. The compartment holds two extra-large, Toronto-type mercury arc lamps. Figure 22 shows the Cary Model 81 with the gas cell unit.

**Liquid Sampling.** Liquid samples can be examined with the cells shown in Fig. 23. Each cell requires a special cell optics assembly and a cell holder. The use of the 19-mm cell has an advantage in that the monochromator does not see the cell walls, and there is a reduction of background and cell fluorescence. As is illustrated in Table III, a sample as small as 0.2 ml can be examined.

**Solid Sampling.** The sampling problem with solids has been one of the most difficult problems to solve in Raman spectroscopy. This appears to be a paradox since some of the early work of Indian investigators was on solids. However, as one works with solids it becomes apparent that one is never entirely successful in obtaining good solid

### TABLE III
### Liquid Cells for the Cary 81

| Cell OD (mm) | Volume (ml) |
|---|---|
| 19 | 65 |
| 7 | 5 |
| 2 | 0.6 and 0.2 |

Fig. 24. Some solid sample tubes for Raman spectroscopy.

Raman spectra. In the first place, with photographic recording instruments, one has to expose the sample for many hours. In the second place, the solid will scatter the exciting line along with the Raman scattering. Turbid samples shield the inner surfaces from the exciting radiation, thus lowering the Raman scattering. Tobin[25] has discussed the problems involved in the Raman spectroscopy of solid materials. Many advances have been made to eliminate these problems. The use of filters to remove the continuum from the mercury source has been reported.[26–29] Advances made by use of a double monochromator have also helped.[30] The use of metal interference filters to transmit the mercury 4358-Å line but reflect the Raman spectrum have also been proposed.[27–29] However, the advent of the photoelectric spectrophotometers did the most to stimulate interest in solid Raman spectroscopy.

New solid sampling techniques have been recently discussed by Ferraro et al.[31] in relation to use with the Cary 81 spectrophotometer. The factors involving the sample tube, sample crystal size, sample thickness, sample position, and sample preparation were considered. Some of the different Raman tubes tried are shown in Fig. 24. Figure 25 shows the solid sample holder manufactured by the Applied Physics Company. Actually, the 19-mm tube, 457 mm in length and with an

Fig. 25. Solid sample holder with KBr pellet.

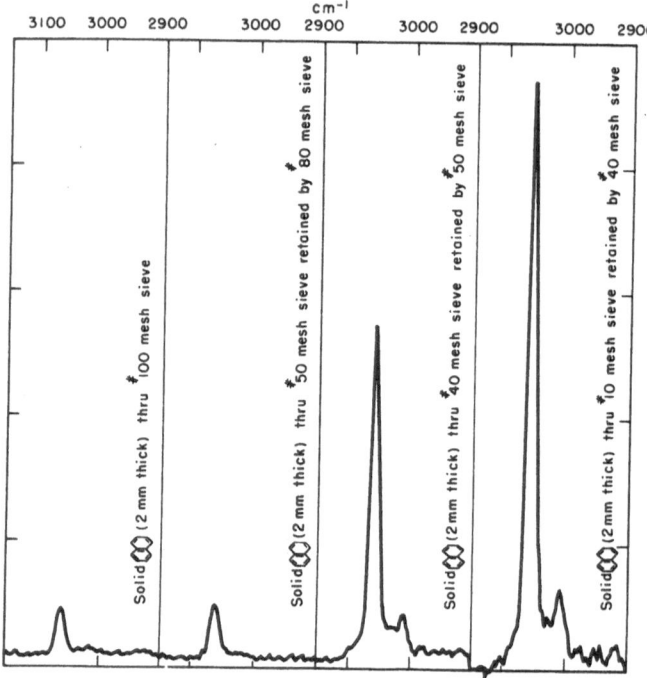

Fig. 26. The effect of crystal size on Raman spectrum of naphthalene.

optical flat Pyrex window at the end, is the most satisfactory from the standpoint of a lowered background and mercury continuum. The effects of crystal size and thickness are illustrated in Figs. 26 and 27, respectively. Placing the sample at the end of the tube is the most versatile method for all of the tubes used. Generally, polycrystalline solids are satisfactory. However, single crystals scatter the best, while

## TABLE IV
### Raman Scattering with Type of Crystal

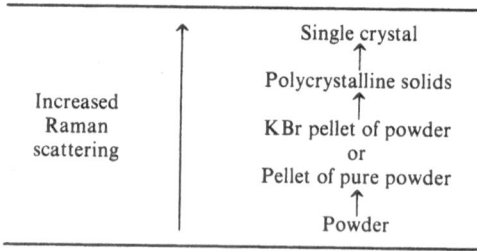

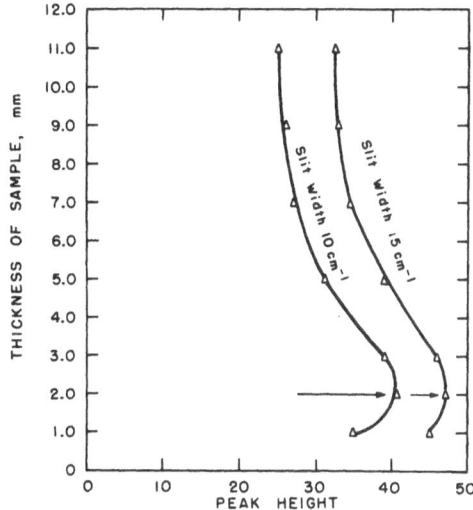

Fig. 27. The effect of sample thickness on Raman
spectrum of naphthalene (region of 3057.5 cm$^{-1}$).

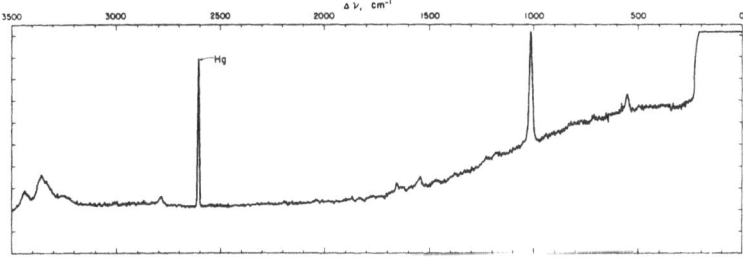

Fig. 28. Solid Raman spectrum of urea (19-mm tube).

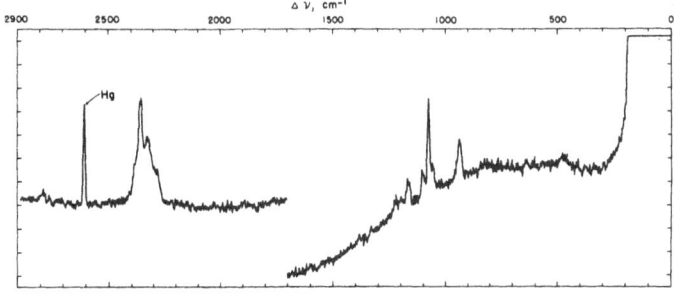

Fig. 29. Solid Raman spectrum of NaH$_2$PO$_2$ (19-mm tube).

fine powders are poor (see Table IV). In the case of polymers, solid rods can be used. Several typical solid spectra are shown in Figs. 28–32.

The use of KBr pellets for solid sampling with photographic instruments was first used by Schrader and co-workers.[32] The KBr pellet technique was first tried with the Cary 81 by Ferraro.[33] The KBr method is suitable for use with fine powders. Comparisons of spectra of powders *versus* KBr disks are shown in Figs. 33–37. Wherever possible, pellets of the material itself are also satisfactory. The pellets can be made as in infrared methods. However, for

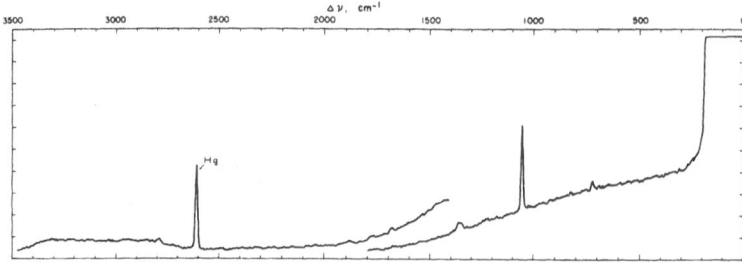

Fig. 30. Solid Raman spectrum of $KNO_3$ (19-mm tube).

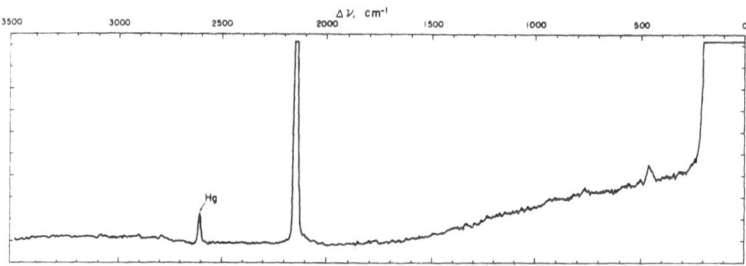

Fig. 31. Solid Raman spectrum of $Cd(SCN)_2$ (19-mm tube).

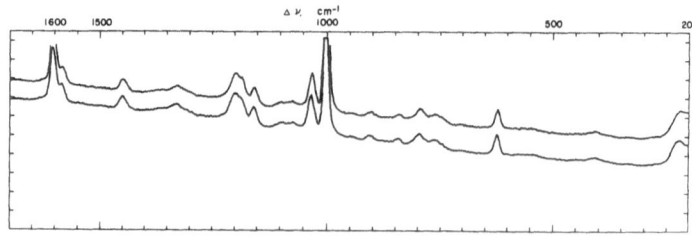

Fig. 32. Solid Raman spectrum of polystyrene rod (8-mm rod).

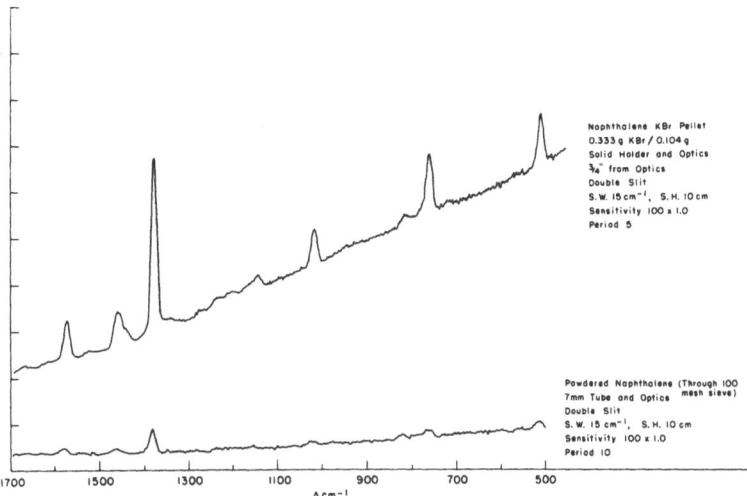

Fig. 33. Comparison of naphthalene pellet and powder spectra.

Raman work with KBr pellets, the pellet must be more concentrated. Concentrations of 5% or more are needed, depending on the scattering properties of the material to be studied. Figures 38 and 39 show the effect of concentration on Raman intensity. The pellets are placed about 1 in. from the solid optics for a better signal-to-noise ratio.

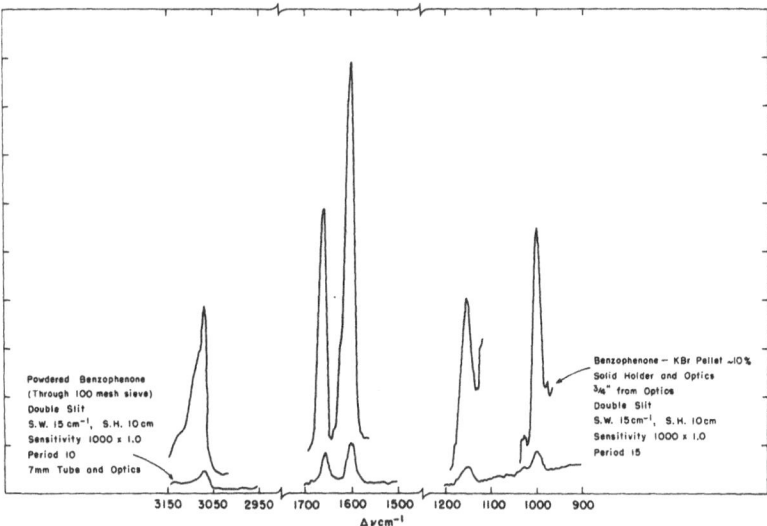

Fig. 34. Comparison of benzophenone pellet and powder spectra.

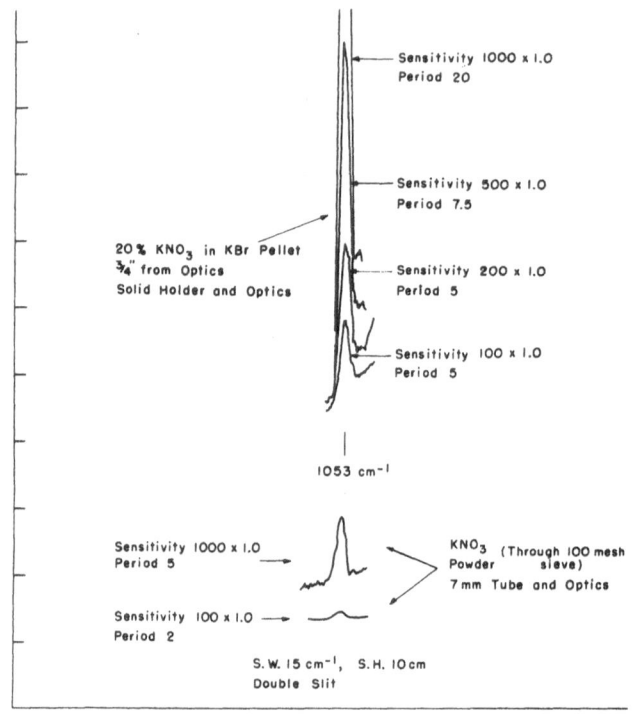

Fig. 35. Comparison of $KNO_3$ pellet and powder spectra.

However, for bands close to the exciting line, Mitchell and Nelson[34] have placed the pellet very close to the solid optics. Figure 40 is the spectrum of naphthalene in KBr placed very close to the solid optics.

**Special Cells.** Recently, Tunnicliff and Jones[35] introduced a special multiple-pass Raman cell for liquid samples for use in a Cary 81, which gives a fivefold increase in the observed Raman intensity. The cell takes advantage of total internal reflections at the cell walls as well as reflections at mirrors at each end of the cell. Diagrams of the multiple-pass cell are shown in Fig. 41.

Several low-temperature cells have been designed for use with the Cary 81. Ferraro and co-workers[36] designed a cell which can be operated in a range of temperatures extending to $-160°C$. The cell is constructed of Pyrex glass 25.5 mm in diameter and can be used with the 19-mm optics of the Cary 81. The cell consists of an inner sample chamber, an intermediate chamber, and an outer jacket which

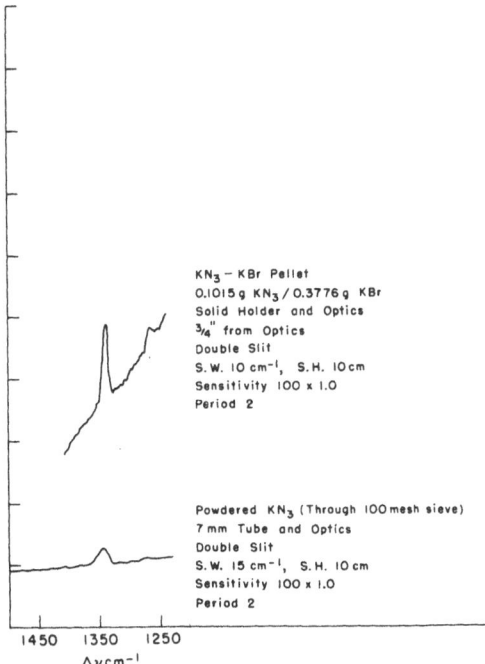

KN₃ − KBr Pellet
0.1015 g KN₃ / 0.3776 g KBr
Solid Holder and Optics
¾" from Optics
Double Slit
S.W. 10 cm⁻¹, S.H. 10 cm
Sensitivity 100 x 1.0
Period 2

Powdered KN₃ (Through 100 mesh sieve)
7 mm Tube and Optics
Double Slit
S.W. 15 cm⁻¹, S.H. 10 cm
Sensitivity 100 x 1.0
Period 2

Fig. 36. Comparison of $KN_3$ pellet and powder spectra.

is evacuated to less than $10^{-4}$ mm Hg. The cooling is provided by helium or nitrogen gas, which is passed through a copper coil immersed in liquid nitrogen. Figure 42 shows a schematic of the cell and Fig. 43 is a photograph of the cell. Griffiths[37] also designed a low-temperature cell. The design, which is shown in Fig. 44, is made of an unsilvered Dewar form. The end tapers to fit snugly into the circular core of the Cary Model 81. The extreme tapered end has parallel optical flats to minimize distortion of the scattering light and reflection losses. The sample tube was made from 7-mm OD Pyrex tubing. A bend was placed in the tube to prevent the liquid from escaping the cooled section of the cell. Samples were condensed in the tube in vacuum, sealed, and stored at $-78°C$ until use. Cooling was provided in the same fashion as described by Ferraro and co-workers.[36]

The use of silvered mirror reflectors with the Applied Physics solid holder (Fig. 45) was introduced recently.[33] The effect of the reflector is to increase the signal-to-noise ratio by 25%. Figure 46 shows the results with naphthalene.

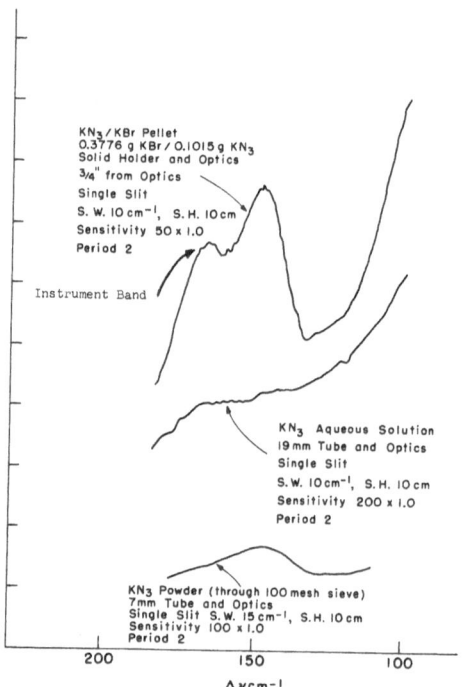

Fig. 37. Comparison of lattice vibration for $KN_3$
pellet, powder, and aqueous solution spectra.

A modification of the Cary 81 for high-temperature studies has been made by Walrafen.[38] A vertical lamp housing containing four vertical arc lamps was mounted on the Raman instrument. An Industrial Instrument A 1-A variable 100-V 100-A DC supply was used to power the mercury lamps. Radio-frequency heating was used, accomplished with a Lepel T-10-3 HB generator, a spiral heating coil, and a slotted cylindrical susceptor of Inconel in contact with a quartz Raman tube. Temperatures were measured with a Chromel–Alumel thermocouple. A schematic diagram of the apparatus is shown in Fig. 47.

## Recent Developments in High-Power Output Lasers for Possible Raman Use

A recent advance in laser science has been the development of 1-W power output lasers.[39,40] These are neon, krypton, or argon gas

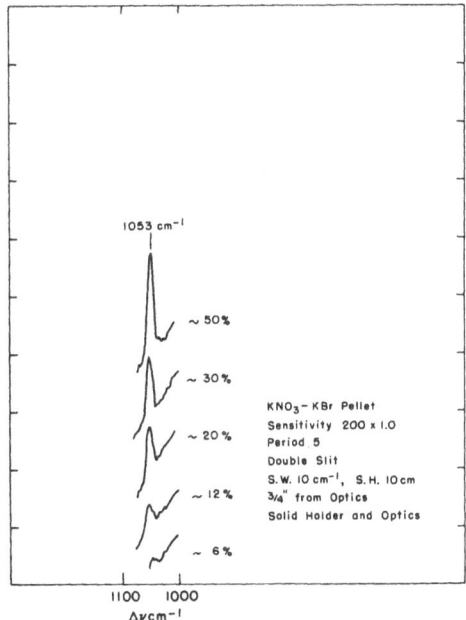

Fig. 38. Effect of $KNO_3$ concentration in KBr
pellet *versus* Raman intensity (1053 cm$^{-1}$ band).

lasers. The argon-ion gas laser is one that is receiving much attention. Porto has recently used such a laser as a Raman source.[41]

Since the progress of laser Raman instrumentation is coupled with the development of better lasers it seems appropriate to discuss the argon-ion laser further. The Raytheon Company manufactures an argon-ion laser (LG12). The laser tube is a DC excited gas discharge tube filled with argon gas at a pressure of about 0.5 mm Hg. The tube is terminated by windows placed at Brewster's angle to reduce light absorption. Two highly reflecting mirrors are placed external to the windows to form the optical resonator. Laser light is produced if the cavity of the resonator is sufficiently high and gain is produced by an inversion of the population of the excited states of the argon atom. The argon-ion laser can be considered as either a three- or four-level system depending on whether the ionic or normal ground state is taken as the lowest level. The most probable transitions for this laser have energies corresponding to wavelengths of 4880, 5145, 4765, 4915, 5017, 4579, and 4658 Å in the order of decreasing probability. The laser is a high-power, low-divergence light source capable of producing

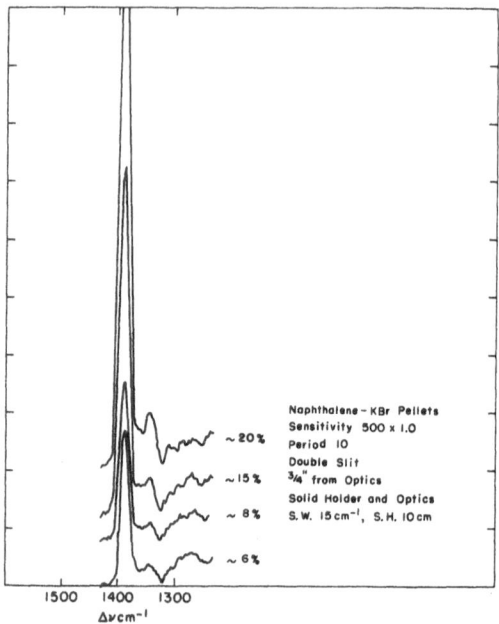

Fig. 39. Effect of naphthalene concentration in KBr
pellet *versus* Raman intensity (1380 cm$^{-1}$ band).

continuous output in either a multimode or single transverse mode
pattern in six wavelengths (4579, 4658, 4765, 4880, 4915, and 5145 Å)
which are part of the blue-green portion of the visible electromagnetic
spectrum. The principal wavelengths are the 4880-Å line at a minimum
power output of 400 mW, the 5145-Å line at a minimum power output
of 250 mW, and the 4765-Å line at minimum power output of 120 mW.
Figure 48 shows the laser with its power supply unit. Figure 49
illustrates the LG12 argon-ion laser in operation, where by the use of
a prism (internal or external to the resonant cavity) the lines from
4579 to 5145 Å may be selected. Further research with such lasers is
necessary to ascertain the laser lifetime *versus* the power output used.
For Raman use it would be necessary that the laser source have a
lifetime comparable to the Toronto mercury arc.

### Perkin-Elmer Laser Raman Spectrophotometer

**Liquid Sampling.** The cell used for liquid sampling with the
Perkin-Elmer laser Raman spectrophotometer is a special design to

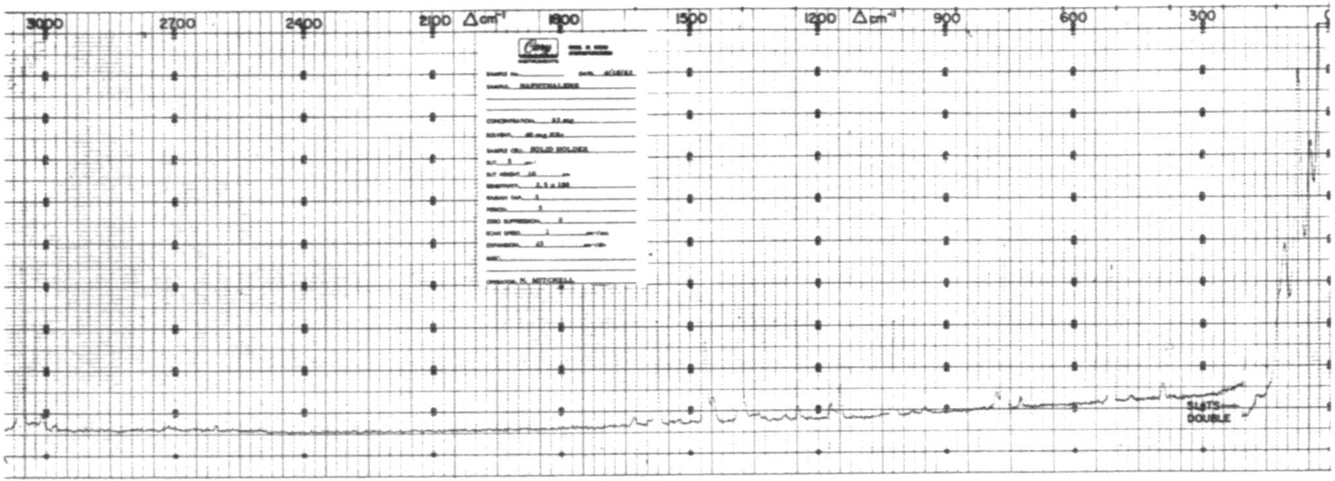

Fig. 40. Spectrum of KBr pellet of naphthalene. Pellet placed very close to solid optics.

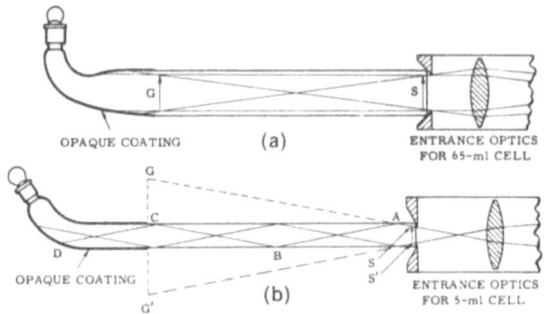

Fig. 41. Multipass liquid cells. (a) Schematic cell optics for 65-ml cell. (b) Schematic cell optics for 5-ml cell.

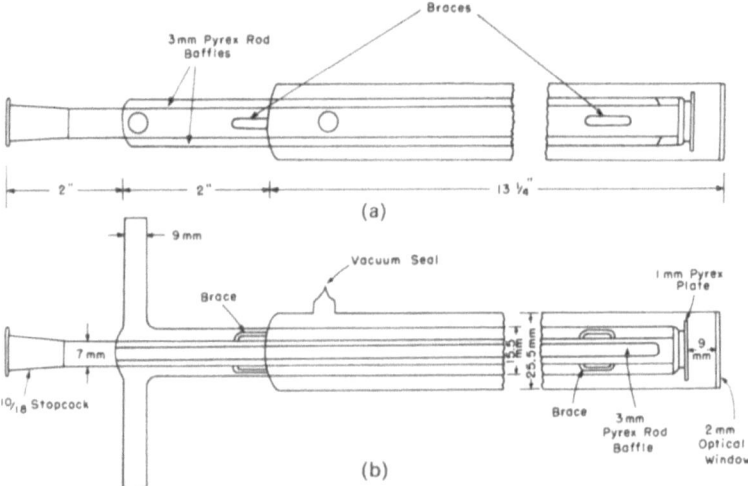

Fig. 42. Schematics of low-temperature 25.5-mm Raman cell of Ferraro and co-workers. (a) Top view. (b) Front view.

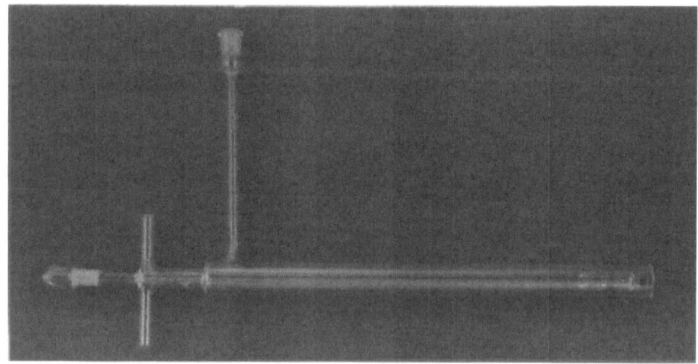

Fig. 43. Photograph of low-temperature Raman cell.

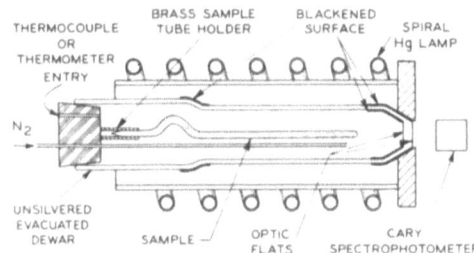

Fig. 44. Low-temperature Raman cell of Griffiths
and co-workers.

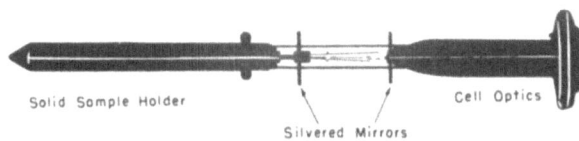

Fig. 45. Solid sample holder with silvered mirrors.

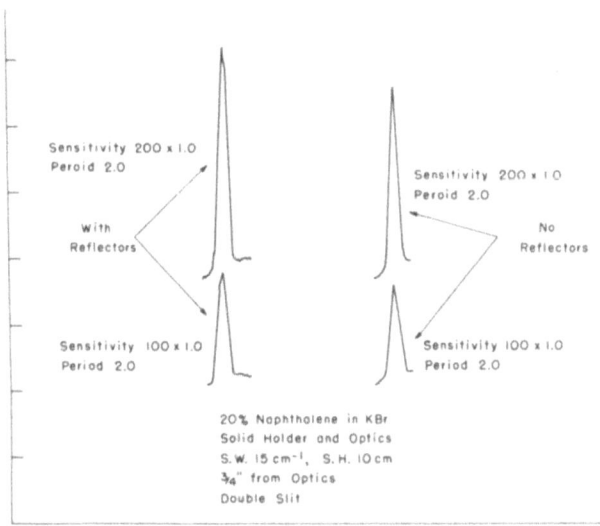

Fig. 46. Naphthalene pellet spectra of 1380 cm$^{-1}$ band
with and without reflectors.

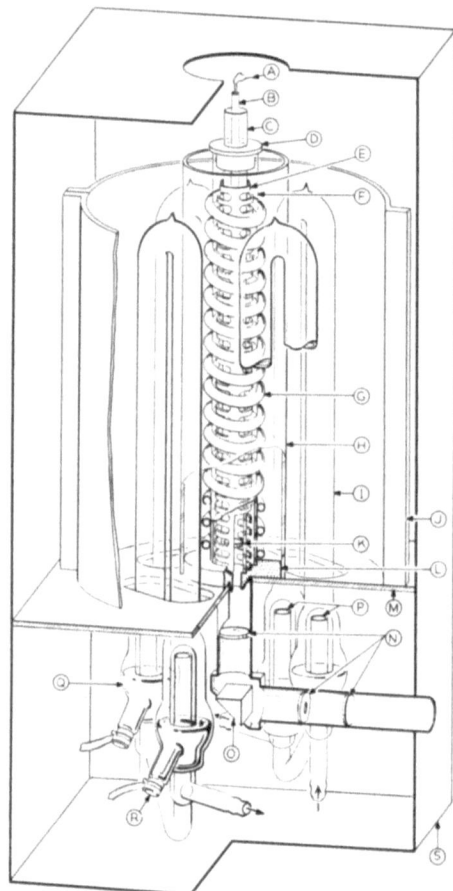

Fig. 47. Schematic of high-temperature accessory for Cary 81. (A) Chromel–Alumel thermocouple; (B) quartz thermocouple well; (C) quartz Raman tube; (D) transite spacer; (E) Inconel susceptor; (F) quartz coil holder and heat shield; (G) copper, water-cooled, radio-frequency heating coil; (H) Vycor heat shield; (I) Pyrex Raman lamp; (J) brass, MgO-coated reflector; (K) quartz optic flat; (L) transite spacer; (M) transite shelf; (N) Cary Model 81 optics and mask; (O) 45° prism; (P) cold fingers; (Q) mercury; (R) Kovar-cup electrode; (S) aluminum lamp housing.

produce multiple passes through the sample. It is a wedged cell with the top and bottom windows coated with a multilayer of dielectric reflective material. The beam is directed vertically to the top of the cell and as many as 150 passes through the sample can occur. In the rear of the cell a collecting spherical mirror gathers the back-scattered

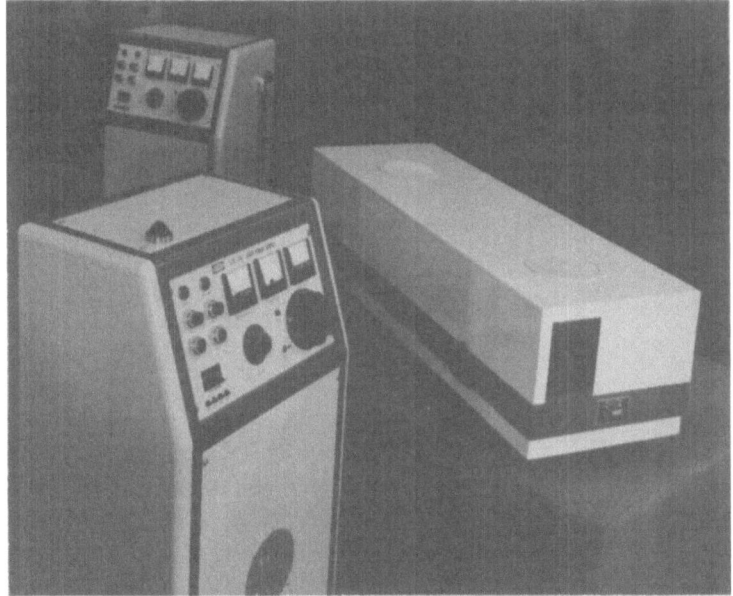

Fig. 48. Raytheon Co. LG12 argon-ion laser with power supply.

Raman energy and focuses it back toward the monochromator entrance slit. Figure 50 shows the new fused silica sample cell (2.5 ml) and the mounting assembly. Figure 51 illustrates the small-volume cell (0.2 ml) in the sample area.

## POSSIBLE FUTURE DEVELOPMENTS

The future will see additional research toward the goal of more powerful monochromatic laser Raman sources. In all probability, new lasers, suitable for Raman use, will be developed. These lasers will be pulsed continuously and emit a more powerful beam. In addition, they will be made to emit exciting energy in the region of the electromagnetic spectrum where one chooses.

Because of these new, more powerful laser Raman sources, the Raman spectra of solids will be easier to obtain. Colored materials will be routinely run in laser Raman spectrometers. Smaller sample sizes can be used. It is anticipated that with a single Raman instrument, without any major instrument modifications (exclusive of the sampling), Raman spectra of gases, liquids, and solids, colored or colorless, will

Fig. 49. Raytheon Co. LG12 laser in operation.

Fig. 50. Laser Raman liquid sample cell (2.5 ml).

Fig. 51.  Laser Raman liquid sample cell (0.2 ml).

be easily obtained.  The expected improvement of the fluorescence problem with a laser source will allow one to study many colored aqueous and fluorescing biological solutions that cannot be adequately studied by infrared techniques.  Complex chemistry of colored aqueous solutions of transition metals can now be studied by Raman techniques. Low-energy transitions close to the exciting line should be easier to see with a laser source.  The cost of laser Raman instruments will be competitive with the high-resolution infrared instruments.  All these factors will serve to develop more interest in Raman spectroscopy in the years to come.

## ACKNOWLEDGMENTS

The author wishes to express his sincerest thanks and appreciation to the Applied Physics Corporation for Figs. 3–7 and 40 relating to the Cary 81; to the Perkin-Elmer Corporation for Figs. 8–20 and

50 and 51 relating to the laser Raman spectrophotometer; to Drs. D. D. Tunnicliff and A. C. Jones for their figure on the multipass Raman cell, and Dr. J. E. Griffiths for the figure on his low-temperature cell; to the Hilger and Watts Company for Figs. 1 and 2; to Dr. George E. Walrafen for the figure on the high-temperature accessory; and to Raytheon Company for their figures on the argon-ion laser. This paper is based on work performed under the auspices of the U.S. Atomic Energy Commission.

## REFERENCES

1. P. Krishnamurti, *Indian J. Phys.* **5**: 587 (1930).
2. R. M. Hoffman and F. Daniels, *J. Am. Chem. Soc.* **54**: 4226 (1932).
3. L. J. Buttolph, *Rev. Sci. Instr.* **1**: 650 (1930).
4. F. P. Kerschbaum, *Z. Instrumentenk.* **34**: 43 (1914).
5. B. Veskatesachar and L. Sibaiya, *Indian J. Phys.* **5**: 747 (1930).
6. J. H. Hibben, *The Raman Effect and Its Chemical Application*, Reinhold Publishing Corp., New York, 1939.
7. F. H. Spedding and R. F. Stamm, *J. Chem. Phys.* **10**: 176 (1942).
8. D. H. Rank and J. S. McCartney, *J. Opt. Soc. Am.* **38**: 279 (1948).
9. H. L. Welsh, M. F. Crawford, T. R. Thomas, and G. R. Love, *Can. J. Phys.* **30**: 577 (1952).
10. N. S. Ham and A. Walsh, *Spectrochim. Acta* **12**: 88 (1958).
11. J. Brandmüller and H. Moser, *Emführung in Die Raman Spektroskopie*, Dr. Dietrich Steinkopff Verlag, Darmstadt, 1962.
12. D. H. Rank and R. V. Wiegand, *J. Opt. Soc. Am.* **36**: 325 (1946).
13. D. H. Rank, R. J. Pfister, and P. D. Coleman, *J. Opt. Soc. Am.* **32**: 390 (1942).
14. J. J. Heigl, B. F. Dudenbostel, J. F. Blank, and J. A. Wilson, *Anal. Chem.* **22**: 154 (1950).
15. R. F. Stamm, C. F. Salzman, and T. Mariner, *J. Opt. Soc. Am.* **43**: 119 (1953).
16. R. F. Stamm, *Ind. Eng. Chem. (Anal. Edition)* **17**: 3181 (1945).
17. R. F. Stamm and C. F. Salzman, *J. Opt. Soc. Am.* **43**: 126 (1953).
18. I. S. Bowen, *Astrophys.* **88**: 113 (1938).
19. W. Benesch and J. Strong, *J. Opt. Soc. Am.* **41**: 252 (1951).
20. J. V. White, *J. Opt. Soc. Am.* **41**: 732 (1952).
21. H. Cary, W. S. Galloway, and K. P. George, paper presented at Molecular Spectroscopy Symposium, Ohio State University, Columbus, Ohio (1953).
22. S. P. S. Porto and D. L. Wood, *J. Opt. Soc. Am.* **52**: 251 (1962).
23. H. Kogelnik and S. P. S. Porto, *J. Opt. Soc. Am.* **53**: 1446 (1963).
24. E. H. Siegler, C. D. Hinman, and A. F. Slomba, paper presented at Molecular Spectroscopy Symposium, Ohio State University, Columbus, Ohio (1964).
25. M. C. Tobin, "Sample Problem in Raman Spectroscopy," in: W. D. Ashby (ed.), *Developments in Applied Spectroscopy*, Vol. 1, Plenum Press, New York, 1962, pp. 205–214.
26. M. F. J. Taboury, *Bull. Soc. Chim.* **10**: 205 (1943).
27. J. Brandmüller, *Z. Angew. Phys.* **5**: 95 (1953).
28. A. Simon, H. Kriegsmann, and E. Steger, *Z. Physik. Chem. (Leipzig)* **205**: 181 (1956).
29. B. Schrader, F. Neidel, and G. Krezge, *Z. Physik. Chem. (Frankfurt)* **12**: 132 (1957).

30. K. W. F. Kohlrausch and A. W. Reitz, *Ramanspektrem,* Edwards Brothers, Inc., Ann Arbor, Michigan, 1945.
31. J. R. Ferraro, J. S. Ziomek, and G. Mack, *Spectrochim. Acta* **17**: 802 (1961).
32. B. Schrader, F. Nerdel, and G. Kresze, *Z. Anal. Chem.* **170**: 43 (1959).
33. J. R. Ferraro, *Spectrochim. Acta* **20**: 901 (1964).
34. N. Mitchell and D. C. Nelson, *J. Chem. Phys.* **39**: 1364 (1963).
35. D. D. Tunnicliff and A. C. Jones, *Spectrochim. Acta* **18**: 569 (1962).
36. J. R. Ferraro, J. S. Ziomek, and K. Puckett, *Rev. Sci. Instr.* **35**: 754 (1964).
37. J. E. Griffiths, R. P. Carter, and R. R. Holmes, *J. Chem. Phys.* **41**: 3863 (1964).
38. G. E. Walrafen, *J. Chem. Phys.* **43**: 479 (1965).
39. E. I. Gordon, E. F. Labuda, and W. B. Bridges, *Appl. Phys. Letters* **4**: 178 (1964).
40. W. B. Bridges and A. N. Chester, *Appl. Opt.* **4**: 573 (1965).
41. S. P. S. Porto, paper presented at the Fourth National Meeting of the Society for Applied Spectroscopy, Denver, Colorado (August–September 1965).

## Chapter 3

# Laser Raman Spectroscopy

## J. A. Koningstein

*Carleton University*
*Ottawa, Ontario, Canada*

---

### INTRODUCTION

At present, the low-pressure mercury light source is the most commonly used in Raman spectroscopy. Under normal operating conditions (2.5 kW) it can be calculated that up to 50 W of energy of such a light source is radiated in the 4358 Å line of the mercury spectrum which is frequently used to excite Raman spectra. However, due to the geometry of the mercury light source, only a small fraction of the emitted radiation is utilized in illuminating the sample; in fact, a reasonable estimate is that about 1 W of power in the 4358 Å line is responsible for the excitation of Raman spectra of liquid samples. Even less radiation is available for irradiating solid samples.

Apart from the rather low efficiency of the mercury lamps, it is also true that they do not produce a constant flux of radiation so that special precautions have to be taken if one wants to measure intensities of Raman bands with any accuracy.

In contrast with the properties of the Toronto light source are those of the optical maser or laser beam. The energy here is contained in a very narrow parallel beam which can be manipulated in any direction one wants. The polarization and intensity are all well controlled, while the half-bandwidth of the laser line is, in most cases, smaller than those of the lines of the mercury or any other light source. Consequently, one might expect that the laser beam is probably the most ideal light source for Raman spectroscopy, assuming of course that the power level and the operating frequency are satisfactory. Optical masers or lasers became available only a few years ago and a

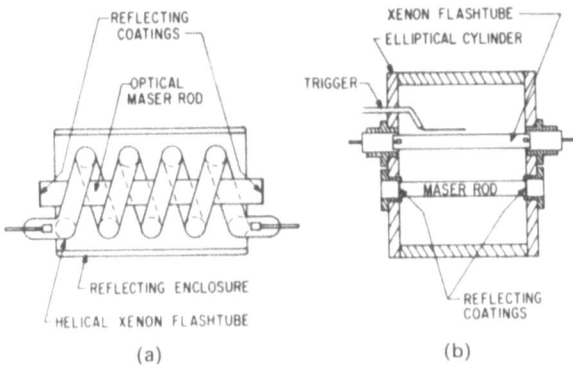

Fig. 1. Two schematic drawings of a ruby maser apparatus. The xenon flash lamp surrounds a ruby rod (a) or is placed in an elliptical reflector (b).

number of Raman experiments have already been performed with these light devices. The possibility of obtaining stimulated emission in the infrared and optical regions was discussed by Schawlow and Townes[1] in 1958. The first successful demonstration of an optical maser was made by Maiman in 1960.[2] He was able to produce a light pulse at 6943 Å during a period of 0.5 msec from a xenon pumped ruby rod (0.05% Cr in $Al_2O_3$). Javan, Bennett, and Herriott[3] developed the continuous-operating He–Ne gas laser in 1961. During the last four years there has been a steady increase in both solid-state and gaseous masers that oscillate at frequencies which extend from the visible and infrared to the microwave spectral regions.[4] Because of their operating frequencies and availability, it has been the pulsed ruby optical maser and the continuous-wave He–Ne gas laser which have been used most widely as light sources in Raman spectroscopy.

From the published experimental results one can draw the conclusion that a considerable amount of attention has been paid to reproduce already-known Raman spectra of some strong scattering liquids, primarily to evaluate the relative merits of the laser and conventional light sources. This chapter contains a discussion of (1) the techniques and results obtained with the pulsed ruby optical maser and the continuous-operating He–Ne gas laser as light sources in Raman spectroscopy, and (2) some Raman experiments in which advantage has been taken of the specific properties of the laser beam.

## GENERAL EXPERIMENTAL TECHNIQUES AND RESULTS

### The Pulsed Ruby Optical Maser as Light Source

As shown in Fig. 1, a typical arrangement of a ruby maser apparatus consists of a ruby rod surrounded by a helical xenon flash lamp. The energy for the lamp is stored in a capacitance bank (several thousand microfarads), and the operating voltage is typically of the order of 2000–4000 V. The output of the maser beam is of the order of joules, and the time duration of the pulse is 0.5 msec. The ruby maser line at 300°K is at 6943 Å.

In 1962 Porto and Wood of the Bell Telephone Laboratories and Stoicheff of the National Research Council of Canada photographed for the first time well-known Raman spectra of some strong scattering organic compounds which were now excited with the (pulsed) ruby maser line. In the experiments of Porto and Wood, the maser beam was focused down in a cell. The walls of this cell were coated with $BaSO_4$ causing the excitation radiation to be diffusely scattered. A simple lens was used to focus the scattered light which emerged from the end of the tube on the slit of a low-dispersion spectrograph. About 50–100 flashes were necessary to photograph weak Raman spectra of $CCl_4$ and $C_6H_6$ on I-N spectrographic film. Stoicheff, on the other hand, employed the principle of multiple reflection in his experiments. The maser beam was allowed to enter a silver-coated

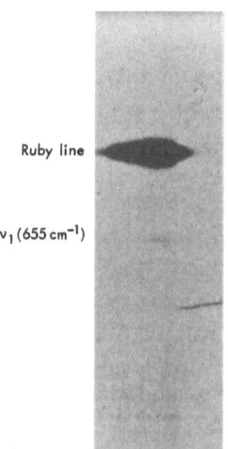

Ruby line

$v_1$ (655 cm$^{-1}$)

Fig. 2. A weak Raman spectrum of $CS_2$ excited with the ruby maser line. The effective exposure time was 20 msec (twenty maser pulses, each lasting about 1 msec).

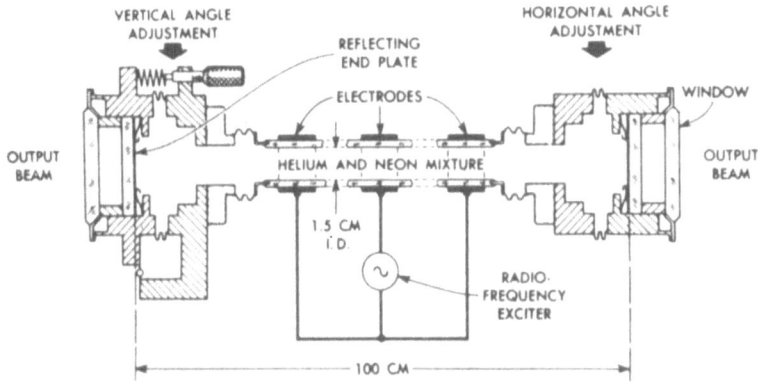

Fig. 3. Schematic of a He–Ne gas laser.

cell (through a small hole in the coating) and was reflected toward the other end of this cell where the excitation radiation was caught in a light trap. Also in these experiments a single lens was used to focus the scattered radiation on the slit of the spectrograph. An effective exposure time of 20 msec was required to photograph the band of $CS_2$ at 655 cm$^{-1}$ (Fig. 2). Although the effective exposure times used to record these spectra were very short, a ruby optical maser can be fired only several times a minute, and the total elapsed time involved to obtain the spectra approaches the time which one needs to photograph similar spectra excited with a low-pressure mercury light source. Nevertheless, the ruby maser could be of importance if the material which is to be investigated absorbs at wavelengths below 6000 Å, where the mercury light source is of no suitable use for exciting Raman spectra.

A radically new phenomenon—the so-called stimulated Raman effect—has been observed if the light of a giant pulsed ruby maser (with powers up to 100 MW and a pulse duration of $10^{-8}$ sec) is introduced in a cell which contains some organic compounds.

**The He–Ne Laser as Light Source**

A schematic illustration of a He–Ne gas laser is given in Fig. 3. The discharge in the tube—filled with a 7:1 mixture of helium and neon gas—is maintained by a stabilized high-voltage power supply

(2000–3000 V). The output power of this device depends most critically on the quality of the multilayer dielectric-coated mirrors. Although stimulated emission from the He–Ne laser can be obtained at a number of frequencies, the most suitable frequency for Raman spectroscopy is the line at 6328 Å because it is highest in intensity and lowest in wavelength. In order to obtain maximum power, one of the mirrors should have a reflection coefficient which is as close as possible to one, with the other mirror transmitting about 0.5% at 6328 Å. The laser tubes which are now most commonly used are between 0.5 and 1.5 m long; they yield powers up to 100 mW in the 6328 Å line. The continuous He–Ne gas laser has been used in a number of Raman experiments.

The power inside the cavity is of course much higher than that outside. Kogelnik and Porto[7] (1963) took advantage of this fact. They placed a cell inside the cavity and with an estimated power of 1.5 W flowing through the liquid in the cell photographed good-quality spectra of $CCl_4$, $C_6H_6$, and $CS_2$ using infrared-sensitive 10-sec-development Polaroid film. The rather weak Stokes line of $CS_2$ at 397 cm$^{-1}$, for instance, could be seen on the photograph but not the overtone of $CCl_4$ which was 1539 cm$^{-1}$ from the exciting line (see Fig. 4). We note in passing that this latter band has been photographed by Schull (1955) in 1 min with a conventional Raman apparatus which employed a mercury light source with standard

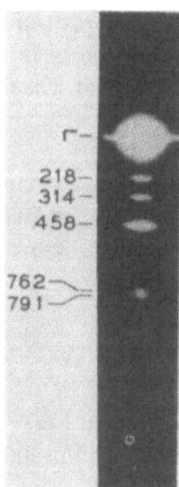

Fig. 4. The Raman spectrum of $CCl_4$ excited with the 6328 Å line of a He–Ne gas laser. The Raman cell was placed inside the laser cavity. Exposure time is 1 hr.

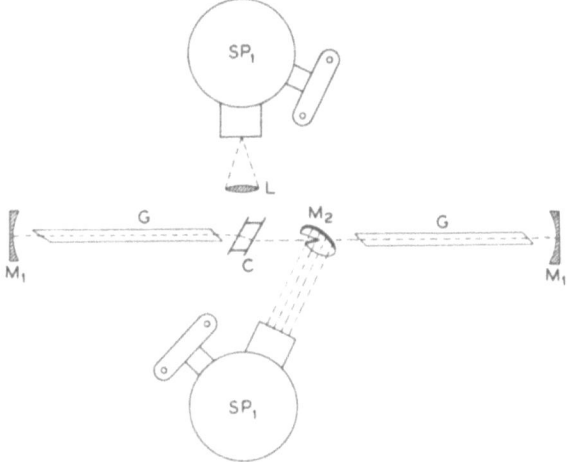

Fig. 5. Experimental arrangement used by Kogelnik and Porto
to observe small angle and 90° angle Raman scattering.

photographic plates. Kogelnik and Porto also reported the observation of Raman scattering which made a small angle with the direction of the incident radiation. In these experiments a slotted mirror was placed behind the Raman cell inside the cavity (Fig. 5). The scattered radiation was collected with this mirror and then focused on the slit of a spectrograph. A 20-sec exposure time was required to photograph the 992 and 3050 cm$^{-1}$ bands of the $C_6H_6$ molecule.

At the beginning of this chapter it was indicated that about 1 W of power in the 4358 Å line of a mercury light source is commonly responsible for the excitation of Raman spectra. The power inside the laser tube is also of the order of 1 W; hence, photoelectric recording of the spectra should, in principle, be possible. Raman intensities, however, are proportional to $v^4$, while the sensitivity of photocathodes for the red spectral region is less than that for the 4000–5000 Å area. Both of these factors would tend to favor the mercury light source. On the other hand, it is true to say that the energy of the laser is contained in a very narrow beam and, therefore, it can be utilized in a more effective way.

Continuous photoelectric recording of Raman spectra excited with the 6328 Å line of the He–Ne gas laser was achieved in 1964 by Leite and Porto[8] and by Koningstein and Smith.[9] Leite and Porto placed the Raman cell inside the laser cavity. The windows of the cell

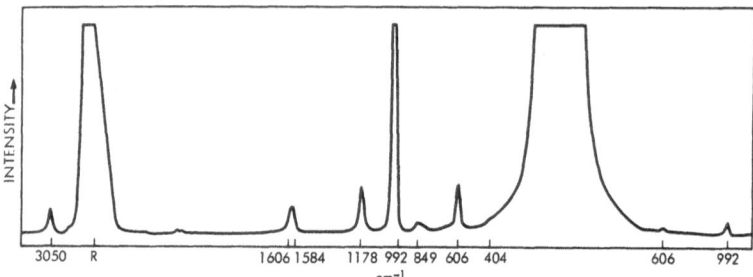

Fig. 6. A photoelectric recorded Raman spectrum of $C_6H_6$ excited with the 6328 Å line of a He–Ne laser. The Raman cell was placed inside the laser cavity.

were set at the Brewster angle in order to avoid cavity losses due to reflection at the cell windows. Also the mirrors of the laser (radius of curvature, 2.5 m) were 3.5 m apart and this geometry yielded a laser beam (in the Raman cell) with a diameter of 0.5 mm. An image of the narrow part of the laser beam was made on the slit of a spectrometer. The detection of the Raman signals was accomplished with cooled EMI 9558B or S-1 Amperox 150 CVP phototubes. A lock-in amplifier was used. The Raman spectrum of $C_6H_6$ recorded under these circumstances with a 12 cm$^{-1}$ slit width and a 0.3 sec time constant is shown in Fig. 6. The authors claimed a signal-to-noise ratio of 3000:1 for the intensity of the $v_1$ vibration of this molecule, which is slightly less than the value which can be obtained with a conventional photoelectric Raman spectrometer. In some instances, especially in the case of solid samples, the scattering losses can be large and it seems preferable to have the Raman cell outside the laser cavity. Under such conditions Koningstein and Smith tried to increase the excitation energy by employing the principle of multiple reflection for the laser beam. They used in their experiments a silver-coated absorption cell. The laser beam was focused and allowed to enter the cell through a small hole in the coating. The Raman scattering was collected with a condensing lens and was imaged on the slit of a 1-m Jarrell-Ash spectrometer equipped with a 600 lines/mm grating blazed for 2.1 μ. The detection of the Raman spectra was accomplished by a cooled EMI 9558B phototube followed by a lock-in amplification. The complicated fine structure of the 655 cm$^{-1}$ band of $CS_2$ recorded in first order with spectral slits of 1.3 cm$^{-1}$ as shown in Fig. 7 may be compared with a photoelectric recording published by Schrötter[10] who employed the conventional technique. Such a comparison reveals that the signal-to-noise ratio of the intensities of the Raman

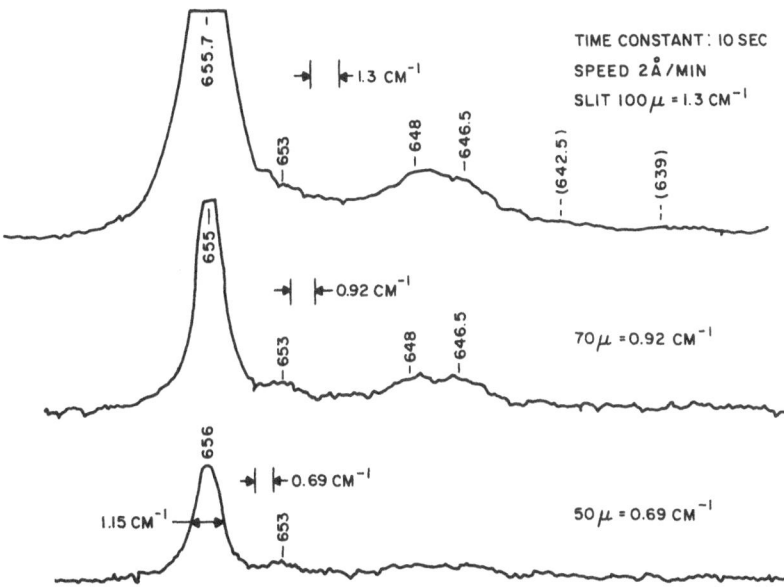

Fig. 7. The fine structure of the 655 cm$^{-1}$ band of $CS_2$. The spectrum was excited with the 6328 Å line of the He–Ne gas laser with the sample cell outside the laser cavity.

bands of $CS_2$ excited with the He–Ne gas laser are down by a factor of about 50.

A commercial laser Raman spectrometer* is now available. Similar to the experiments described above, the Raman cell is placed outside the laser cavity. A dielectric-coated slightly wedged cell is used as the cell for this special geometry causing the laser beam to be reflected in an orderly way in a plane which coincides with the plane in which the slit and the optical axis of the spectrometer are situated. The optical system employs a reflecting mirror behind the cell and a lens to focus the scattered radiation onto the slit of a monochromator. The signals are detected with a noncooled phototube with S-20 response followed by AC amplification and are then recorded. The signal-to-noise ratios for a number of Raman bands of different liquids obtained with this instrument (power of the laser, 5 mW) compare favorably with the signals which have been recorded with conventional Raman spectrometers.

* The Perkin-Elmer Model LR-1 laser-excited Raman spectrometer.

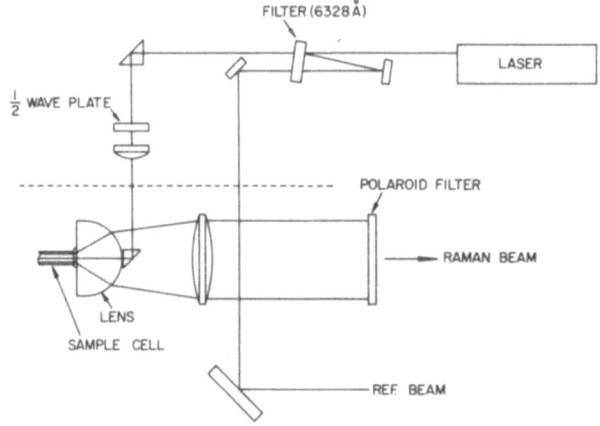

Fig. 8. The optical diagram of the Cary laser excited Raman spectrometer.

Also, the Cary Raman spectrometer from the Applied Physics Corporation is presently being modified so that laser-excited spectra can be recorded. The optical system employed is that shown in Fig. 8. According to presently available information, one expects a performance which compares very well with the 4358 Å mercury-excited spectra of the original spectrometer.

## SPECIFIC LASER RAMAN EXPERIMENTS AND RESULTS

Only a limited number of experiments have been carried out in which advantage has been taken of the intrinsic properties of the laser radiation. These properties have permitted studies of spectral features which were not possible with conventional light sources. Short descriptions of these experiments are given.

The angular dependence of Raman scattering, for instance, had never been studied before, due to the fact that no suitable light source was available for such experiments. Damen et al.[11] were first to report such a study on a number of Raman bands of the benzene molecule in the condensed phase.

In this experiment, the beam of the He–Ne laser made a single pass through a Raman cell. The monochromator which was used to isolate the Raman frequencies was mounted on an arm and could be rotated around the geometrical center of the cell. The angular distribution of the intensity of the 992 cm$^{-1}$ frequency of $C_6H_6$ obtained

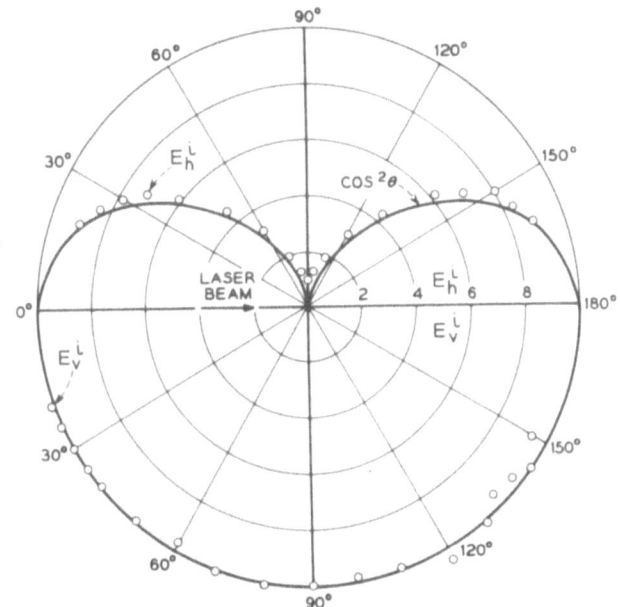

Fig. 9. The angular dependence of the Raman scattered 992 cm$^{-1}$
line of benzene ($E_h^i$ is normalized to 10 at 180°).

in this way is shown in Fig. 9. Both polarizations of the incident
radiation were present. For the incident electric field vector perpendicular to the direction of observation ($E_v^i$), the intensity distribution does not vary with the angle of observation. For the electric field
vector parallel to the observation plane ($E_h^i$), the intensity distribution
follows a cos$^2\theta$ law. These data and similar results obtained for the
intensities of the 1583–1606 and 3049–3062 cm$^{-1}$ doublets of benzene
are important and significant in that they are in agreement with
Placzek's polarizability theory of Raman scattering. The same authors
also reported values of the Rayleigh-to-Raman ratios and absolute
cross sections for the strong Raman lines of $C_6H_6$.

It has become increasingly clear that such Raman intensities
of compounds in the liquid phase are dependent on: (1) a geometric
effect, (2) the internal field effect, and (3) intermolecular interactions.
These effects are by no means small. For instance, the ratio of the
intensities of the 459 cm$^{-1}$ line of the $CCl_4$ molecule in the gaseous
and liquid phases is close to 2.2 instead of 1.[12]

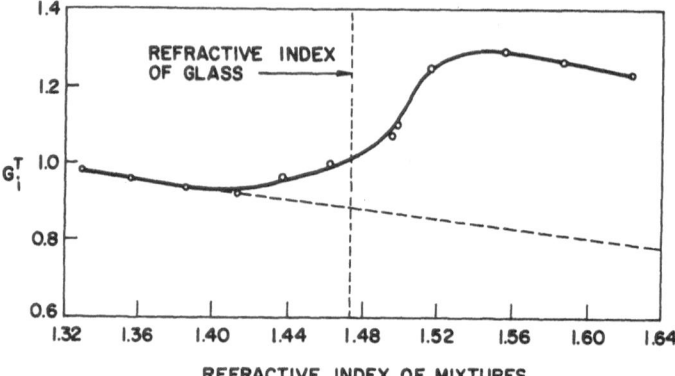

Fig. 10. The influence of the refractive index of the glass of the Raman tube of a conventional spectrometer on Raman intensities of liquids of different refractive index. $G_i^T$ refers to the amount of incident radiation for a conventional Raman apparatus (normalized at 1.0 for $n = n_{CCl_4}$).

## TABLE I
### Depolarization Ratios of Some Raman Bands

| Compound | $\Delta v$(cm$^{-1}$) | (a)* | (b) | (c) | (d) | (e) | $\rho_n^g$† (f) | (g) | $\rho_n^{obs}$ (h) | $\rho_n^L$ (i) |
|---|---|---|---|---|---|---|---|---|---|---|
| CCl₄ | 218 | 0.89 | | | | | | | 0.87 | 0.88 |
| | 314 | 0.88 | | | | | | | 0.88 | 0.85,7‡ |
| | 458 | 0.064 | 0.013 | | | | | | 0.02₆ | 0.010 |
| | 762 | 0.86 | | | | | | | 0.85 | 0.86 |
| | 790 | 0.86 | | | | | | | | |
| CHCl₃ | 262 | 0.86 | | 0.86 | 0.88 | | | | 0.84 | 0.85 |
| | 366 | 0.19 | 0.16 | 0.20 | 0.24 | | | | 0.22₅ | 0.20 |
| | 668 | 0.082 | | 0.06 | 0.05 | | | | 0.04₈ | 0.03₂ |
| | 771 | 0.93 | | 0.86 | 0.8₉ | | | | 0.86 | 0.87 |
| C₆H₆ | 606 | 0.87 | | | | 0.86 | | | 0.82 | 0.83 |
| | 850 | 1.00 | | | | 0.86 | | | | |
| | 992 | 0.087 | 0.055 | | | 0.11 | | 0.02₉ | 0.05 | 0.03₅ |
| | 1175 | 0.83 | | | | 0.86 | | | 0.84 | 0.85 |
| CS₂ | 650 | 0.28 | | | | | 0.2₅ | 0.28₅ | 0.31 | 0.29₄ |
| Cyclohexane | 802 | 0.063 | | | | | | | 0.16₅ | 0.15 |

* Columns labeled (a)–(i) refer to the work of the following authors: (a) Rank;[20] (b) Douglas and Rank;[21] (c) Stamm, Salzman, and Mariner;[22] (d) Cabannes and Rousset;[23] (e) Allen and Bernstein;[24] (f) Evans and Bernstein;[25] (g) Koningstein and Bernstein;[26] (h) Koningstein (this work); (i) corrected values for this work.

† $\rho_n$ denotes the depolarization ratio for natural light. $\rho_n^g$ denotes the depolarization ratio measured according to the method of Edsall and Wilson.[27]

‡ The depolarization ratio of this band was assumed to be $\frac{6}{7}$.

The geometric effect is related to the influence of the refractive index of the liquid on the amount and distribution of the excitation radiation inside the Raman cell and on the amount of Raman scattering which is collected by the optical system and then detected. Different conventional Raman spectrometers employ different illumination arrangements and, hence, result in different geometric effects. This is probably the reason why a rather wide range of values is reported for intensity ratios of Raman lines of the same compounds. Koningstein[13] employed the He–Ne gas laser to measure relative intensities of Raman lines and in doing so obtained information on the geometric effect of a conventional Raman spectrometer. A comparison was made of values of intensities of the $\nu_1$ vibration of $CCl_4$ in mixtures with $CS_2$ and $CH_3OH$ recorded with laser Raman and conventional spectrometers. Although the intensity distribution of the laser radiation in a multiple-pass reflection cell may still be dependent on the refractive index of the liquid in the cell, it can now be said that the refractive index of the glass of the cell does not influence the amount of exciting radiation which is allowed to enter the cell. This is not the case for a Raman tube of a conventional spectrometer. It can be shown that for such a tube an additional amount of exciting laser is allowed to enter the medium if its refractive index becomes larger than that of the glass of the said tube. In comparing then the above-mentioned intensities, the effect of the glass of the Raman tube on intensities of mixtures with different refractive indices became quite apparent and can be quite large (see Fig. 10). The additional amount of incident radiation is different for various conventional arrangements and consequently the values of the intensities of Raman bands should be viewed with great doubts if measured with such apparatus. The same author also measured depolarization ratios of a number of Raman lines of different compounds. In order to obtain an accurate value of $\rho_n$ for the 459 cm$^{-1}$ frequency of $CCl_4$, intensities were measured with the beam of the He–Ne laser making one single pass through the cell. Both directions of polarization of the incident radiation were present, and care was taken to reduce the amount of elliptical polarized light to less than 1 part in $10^5$. A value of $\rho_n$ of $0.010 \pm 0.001$ was obtained for the $\nu_1$ vibration of $CCl_4$ with the single-pass experiment.

Other values of $\rho_n$ were measured with a laser Raman spectrometer which employed a multiple-pass silver-coated Raman cell. The reflections of the laser light at the liquid–glass and glass–silver interfaces are apt to introduce a change in the direction of polarization of the laser radiation. A correction could be made by using the value of $\rho_n$ for the $v_1$ vibration of $CCl_4$. The corrected values of $\rho_n$ are given in Table I together with values which have been obtained with conventional spectrometers.

The excitation of a Raman spectrum of III-V semiconductor with the 6328 Å line of a He–Ne gas laser as reported by Hobden and Russell[14] of the Royal Radar Establishment serves as another example where advantage has been taken of the properties—in this case the frequency—of the laser beam. In the past, no Raman experiments have been made of a III-V semiconductor because of the lack of a suitable light source having an excitation energy less than the band gap. A first- and second-order Raman spectrum could be photographed in 12 hr with an $f/6$ Hilger and Watts Raman spectrograph using a 20-$\mu$ slit width. A tentative assignment of these spectra was given, which awaits confirmation by the results of low-temperature and polarization studies on a single crystal.

The Raman spectrum of rutile, as shown in Fig. 11 at room temperature and at 4.2°K as recorded by Koningstein and Singh, demonstrates the feasibility of low-temperature studies of vibrational spectra. The sample, a single-crystal 4-mm square of 2 cm depth, was mounted on the tip of a cold finger of a helium research Dewar. The laser beam

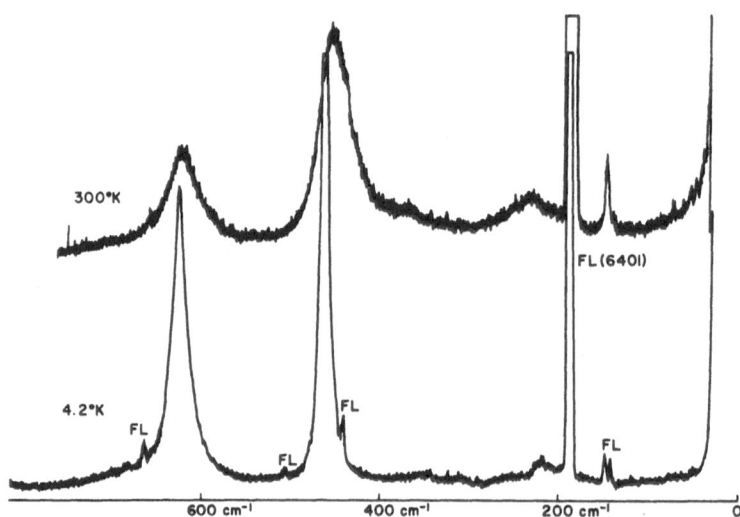

Fig. 11. The Raman spectrum of rutile ($TiO_2$) at different temperatures excited with the He–Ne laser. Time constant, 0.5 sec; FL, fluorescence lines.

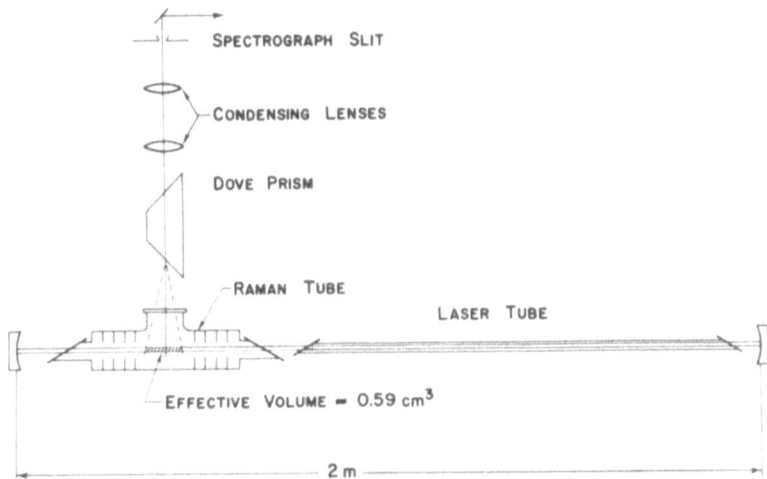

Fig. 12. Experimental apparatus of Weber and Porto employed to excite the Raman spectrum of a gas with the He–Ne laser.

passed through a window in the bottom of the Dewar and was focused down in the crystal. The Raman scattering was observed through another window at right angles.

The pure rotational Raman spectrum of the gas methylacetylene was recently reported by Weber and Porto.[16] A schematic diagram of the experimental apparatus is shown in Fig. 12. The Raman cell was placed inside the laser cavity. The windows of the cell were set at

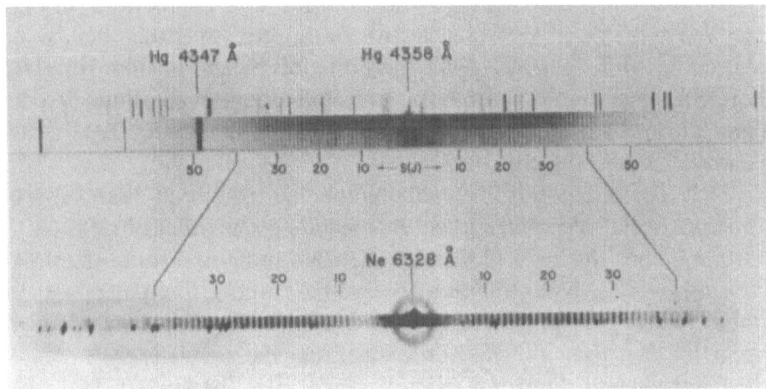

Fig. 13. The pure rotational Raman spectrum of methylacetylene excited with the 4358 Å Hg line and the 6328 Å line of the He–Ne laser.

the Brewster angle in order to reduce the loss of the laser radiation to a minimum value. The volume of the gas which was effectively involved to produce Raman scattering amounted to $0.5\,g/cm^3$; the diameter of the laser beam inside cell was 5 mm and the length of the cell 30 mm. The horizontal distribution of Raman scattering sources was imaged on the vertical slit with the help of a Dove prism. The monochromater was equipped with a 300 lines/mm grating which produced in the ninth order a resolution of $2.0\,cm^{-1}/mm$ on the photographic plate. A cylindrical lens was placed in front of the plate in order to photograph the rotational Raman spectrum in reasonable exposure times. The Raman spectrum shown in Fig. 13 was photographed in 58 hr, the pressure of the gas in the cell being $\frac{1}{2}$ atm. The power of the external laser beam measured with mirrors of 99.9 and 99.5% reflectively was 20 mW. In the actual experiment, two mirrors were used with reflectivities of 99.9% each. The authors compared the results of these laser experiments with data on the excitation of the Raman spectrum with the 4358 Å line of a mercury light source, which is also shown in Fig. 13.

The ultimate role of the laser in Raman spectroscopy is not only that of a convenient light source, but also that of a convenient and *extremely monochromatic* light source. This means, for example, that ideally the He–Ne laser should be made to operate not only in just one band, for example, the 6328 Å band, but also in only one cavity mode in that band. One finds, however, in practice that stability and narrow line widths are much more easily obtained over short periods of time (1 sec) than over extended periods of time. This means that, in high-resolution work, intensity will still be a problem—one that is not overcome with conventional methods of detection involving, for example, accumulation of data over long periods of time. In this connection, two detection methods devised by Pao, Zittert, and Griffith[17] are of interest.

The first method is for resolving the fine structure of band spectra and appropriate spectral slit widths used are of the order of $1$–$0.1\,cm^{-1}$. At first glance the method seems to depend upon discarding the slowly varying or DC signal, making use of the random fluctuations in the photodetector output instead. The basis for its operation depends upon the fact that information concerning the signal is contained in the "noise" and, for some circumstances, it is better to obtain this information in a seemingly indirect way. The experimental arrangement used by Pao, Zittert, and Griffith is that shown schematically in

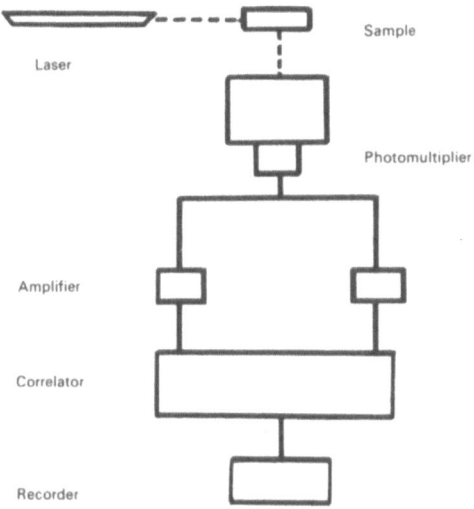

Fig. 14. Block diagram of the detection system used by Pao, Zittert, and Griffith.

Fig. 14. The output of the amplifiers is passed through a filter to remove the DC signal. The randomly fluctuating components are passed through two parallel amplifying paths and are then multiplied together and integrated to yield a DC output proportional to the initial

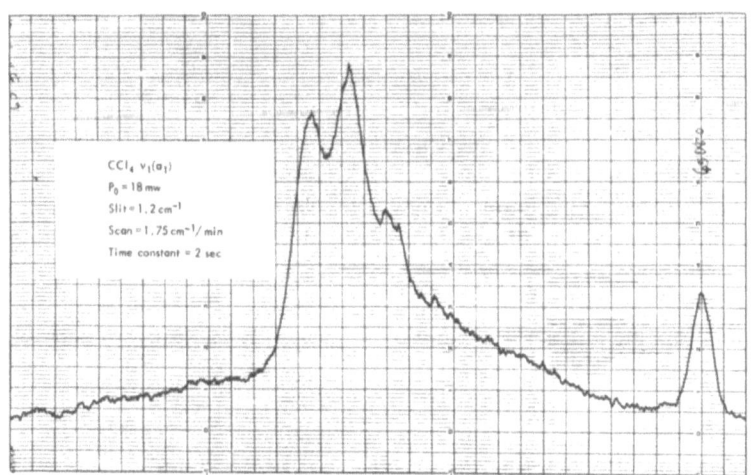

Fig. 15. The isotope fine structure of the 459 cm$^{-1}$ line of CCl$_4$. The Raman spectrum was excited with the 6328 Å line of the laser.

DC signal but with negligible fluctuation. As the setting of the mono-chromator is changed, the level of the DC output also changes and the Raman spectrum is thus scanned. Because the incident line is very narrow, the line shapes so determined will be the true line shapes rather than those partially obscured or averaged by the line width of the incident radiation. Since this method is suitable for operation at very low light intensities, the spectral slit width may be made quite small and high resolution is obtained. The isotope fine structure of the 458 cm$^{-1}$ line of $CCl_4$ was studied with this method and the spectrum so obtained (Fig. 15) is a factor of 30 better (signal-to-noise ratio) than that obtained by the same authors with phase-sensitive amplification methods. This recording is at least comparable with and perhaps better than results obtained with conventional techniques. The validity of this method depends upon the fact that even when the incident radiation is so monochromatic that beats can be neglected, the output of the photodetector is not constant but fluctuates due to the quantum and random nature of photoelectron emission. The power spectrum of this *shot noise*[18] is essentially that of white noise except that there is a cutoff at a frequency prescribed by the RC characteristics of the photo-detector circuit. We note that with band spectra, the Raman scattered radiation incident on the photodetector is not monochromatic, and the instantaneous light intensity is correspondingly not constant but contains fluctuations due to beats between all the frequencies contained within the spectral slit width. These fluctuations are often referred to as *photon noise*. However, it may be shown[19] that in the optical region, for the circumstances just described, the current fluctuations due to photon noise are about $10^{-2}$ that of shot noise and consequently may be neglected. Under such conditions it is appropriate to take a portion of the shot-noise power proportional to $N^{\frac{1}{2}}$ where $N$ is the average emission rate in terms of photoelectrons/sec, separate it into two paths, amplify both currents, and multiply to obtain a signal proportional to $N$, the original signal. This method apparently works very well for weak signals.

For line spectra, the Raman scattered radiation cannot be taken to be of constant amplitude across the bandwidth of the mono-chromator and indeed the photon-noise power is not necessarily small with respect to shot-noise power. Under these circumstances it is essential to remove the shot noise by cross correlation of the signals from two photodetectors looking at the same signal. The beats are correlated but the shot-noise fluctuations are not correlated. In this latter method for the detection of line spectra, an extremely stable local oscillator may be used to shift the beat frequencies, which may be as high as 1 kMc, to lower values more suitable for correlation.

## CONCLUDING REMARKS

Presently it can be said that a start has been made with the introduction of optical masers in Raman spectroscopy. The results reported by Pao, Zittert, and Griffith indicate that with the He–Ne laser as excitation source, spectra can be produced which are of the same quality as spectra which have been recorded photoelectrically with conventional Raman spectrometers.

A CW optical maser which oscillates frequencies around 4900 Å—the argon laser—is presently operational. Although its lifetime is still rather short, the power delivered in these lines can be as high as 500 mW–1 W. A conservative estimate indicates that with such a light source and with the laser beam making one pass through the sample, Raman spectra can be produced which are much stronger—probably by two orders of magnitude—than those obtainable with conventional techniques.

Another CW optical maser, Nd in YAlG, produces radiation at $1.06 \mu$ with power levels around 1 W. Although this frequency is unfavorable for Raman spectroscopy, there is always the possibility of second-order harmonic generation. This technique then might result in stimulated emission at 5300 Å, which wavelength is perfectly acceptable.

## ACKNOWLEDGMENTS

The author is grateful to Drs. Porto and Stoicheff for making photographs of their spectra and apparatus available. He also wishes to extend his thanks to Drs. Pao and Griffith for supplying the information on the detection system.

## REFERENCES

1. A. L. Schawlow and C. H. Townes, *Phys. Rev.* **112**: 1940 (1958).
2. T. H. Maiman, *Nature* **107**: 493 (1960); T. H. Maiman, R. H. Hoskins, I. J. D'Haenens, C. K. Asawa, and V. Evtuhov, *Phys. Rev.* **123**: 1151 (1961); R. J. Collins, D. F. Nelson, A. L. Schawlow, W. Bond, C. G. B. Garrett, and W. Kaiser, *Phys. Rev. Letters* **5**: 303 (1960).

3. A. Javan, W. R. Bennett, and D. R. Herriott, *Phys. Rev. Letters* **6**: 106 (1961).
4. See, for instance, A. J. Bevols and W. A. Barker, *Appl. Opt.* **4**: 531 (1965).
5. S. P. S. Porto and D. L. Wood, *J. Opt. Soc. Am.* **52**: 251 (1962).
6. B. P. Stoicheff, *Tenth International Spectroscopy Colloquium*, University of Maryland, June 1962, Spartan Books, Washington, 1963.
7. H. Kogelnik and S. P. S. Porto, *J. Opt. Soc. Am.* **53**: 1446 (1963).
8. R. C. C. Leite and S. P. S. Porto, *J. Opt. Soc. Am.* **54**: 9011 (1964).
9. J. A. Koningstein and R. G. Smith, *J. Opt. Soc. Am.* **54**: 1061 (1964).
10. H. W. Schrötter, *Z. Elektrochem.* **64**: 853 (1960).
11. T. C. Damen, R. C. C. Leite, and S. P. S. Porto, *Phys. Rev. Letters* **14**: 1 (1965).
12. H. W. Schrötter and H. J. Bernstein, *J. Mol. Spectry.* **12** (1): 1 (1964).
13. J. A. Koningstein, *Abstracts of Eighth Congress of Molecular Spectroscopy* (Copenhagen), 1965.
14. M. V. Hobden and J. P. Russell, *Phys. Letters* **13**: 39 (1964).
15. J. A. Koningstein and S. Singh (unpublished results).
16. A. Weber and S. P. S. Porto, *J. Opt. Soc. Am.* (to be published).
17. Y.-H. Pao, R. N. Zittert, and J. E. Griffith, Symposium on Molecular Structure and Spectroscopy, Ohio State University, Columbus, Ohio June 1965.
18. S. O. Rice, in: Nelson Wax (ed.), *Noise and Stochastic Processes*, Dover Publications, Inc., New York, 1954.
19. R. Homburg Brown and R. Q. Tivios, *Proc. Roy. Soc. (London)* **242A**: 300 (1957).
20. D. H. Rank, *J. Opt. Soc. Am.* **37**: 789 (1947).
21. A. E. Douglas and D. H. Rank, *J. Opt. Soc. Am.* **38**: 281 (1948).
22. R. F. Stamm, C. F. Salzman, and Th. Mariner, *J. Opt. Soc. Am.* **43**: 119 (1953).
23. J. Cabannes and A. Rousset, *Ann. Chim. Phys.* **19**: 233 (1933).
24. G. Allen and H. J. Bernstein, *Can. J. Chem.* **32**: 1124 (1954).
25. J. C. Evans and H. J. Bernstein, *Can. J. Chem.* **34**: 1127 (1956).
26. J. A. Koningstein and H. J. Bernstein, *Spectrochim. Acta* **18**: 1249 (1962).
27. J. T. Edsall and E. B. Wilson, *J. Chem. Phys.* **6**: 124 (1938).

*Chapter 4*

# Raman Intensities and the Nature of the Chemical Bond

Ronald E. Hester

*University of York*
*York, England*

The experimental determination of Raman intensities is commonly a difficult procedure, particularly when the resulting intensities are to be placed on an absolute scale. For the purpose of evaluating derived properties characteristic of the chemical bonds responsible for the intensities, Raman intensities which have been corrected for such factors as spectrometer spectral sensitivity, optical absorption, sample geometry, and refractive index must be used. However, before discussing these factors and the applications and interpretations of absolute Raman intensities, it is necessary to examine in some detail the various theoretical relationships used to relate such corrected intensities to properties of chemical bonds.

## THEORY OF RAMAN INTENSITIES

In the introductory chapter of this volume, it was shown that a quantum mechanical expression[1] for the intensity of Raman scattering by (nonrotating) molecules may be written as

$$I = N_n \frac{64\pi^2}{3c^2}(v_0 - \Delta v)^4 [P]_{nm}^2 \tag{1}$$

where $N_n$ is the number of molecules in an initial state characterized by a set of vibrational quantum numbers $n$, $v_0$ is the frequency of the incident light, $\Delta v$ is frequency shift on scattering, and $P_{nm}$ is the transition probability linking the initial states of the molecules to the final states, characterized by quantum numbers $m$. The transition probability factor is expressed in terms of the induced dipole moment vector $P$, which in turn is related to the molecular polarizability $\alpha$

through the equation $\mathbf{P} = \alpha\mathbf{E}$, where $\mathbf{E}$ is the electric vector of the incident light. Also in the introductory chapter, the classical Placzek[2] theory was used to obtain expressions for the transition probability components in terms of the first derivatives with respect to the normal coordinates of displacement[3] $Q_i$ of the components of the polarizability tensor $\alpha$. If the restriction that the scattering molecules be non-rotating is removed and if treatment involves only the mean value of the derived polarizability tensor $\bar{\alpha}'(=\partial\bar{\alpha}/\partial Q)$ and its anisotropy $\gamma'$, as defined in Chapter 1, expressions are obtained for the squares of each of the three components of the transition probability $P_{nm}$. Taken with equation (1), these lead to expressions for the intensity of Raman scattering in directions at right angles to that of the incident light, and for the degree of depolarization $\rho$ of this scattered light.

The expressions obtained in the introductory chapter for the components of $[P]_{nm}^2$ may now be used to determine the intensity of scattering in an arbitrary direction making an angle $\theta$ with the direction of irradiation. If the components of $[P]_{nm}$ are resolved at right angles to the arbitrarily chosen direction of scattering and the mean square contributions are added, the following effective value of $[P]_{nm}^2$ is obtained:

$$[P]_{nm}^2 = \frac{(n+1)}{8\pi^2\Delta\nu} \frac{\mathbf{E}^2}{45} \{[45(\bar{\alpha}')^2 + 7(\gamma')^2](1 + \cos^2\theta) + 6(\gamma')^2 \sin^2\theta\} \quad (2)$$

Calculation of the intensity of Raman scattering must take into account the fact that although in general most molecules will be in the lowest vibrational energy level ($n = 0$), a small fraction will be in vibrationally excited states. For these states ($n > 0$), the probability of Raman transitions is higher than that of the ground state, since the transition probability is proportional to $n + 1$. Writing $fr$ for the fraction of molecules with $n = r$, it is clear that the Raman intensity will be proportional to

$$\sum_r fr(r + 1)$$

in addition to the total number of molecules present (i.e., to the molar concentration $M$). With use of the relation

$$\sum_r fr(r + 1) = \left[1 - \exp\left(\frac{-h\Delta\nu}{kT}\right)\right]^{-1}$$

derived in Chapter 1, together with equations (1) and (2), it is seen that the intensity $I_\theta$ of Raman scattering in the direction $\theta$ will be

$$I_\theta = \frac{KI_0 M(\nu_0 - \Delta\nu)^4}{\Delta\nu[1 - \exp(-h\Delta\nu/kT)]}$$
$$\times \{[45(\bar{\alpha}')^2 + 7(\gamma')^2](1 + \cos^2\theta) + 6(\gamma')^2 \sin^2\theta\} \qquad (3)$$

where $I_0$ (proportional to $\mathbf{E}^2$) is the incident light intensity, and $K$ is a constant.

It is necessary to consider now the possible experimental conditions under which measurements of relative intensities may be interpreted meaningfully. The conventional Wood's tube[4] arrangement of the sample relative to the light source employs a range of angles ($\theta$) between the directions of irradiation and scattering. The problem has been treated by Woodward and Long.[5] Replacement of $I_0$ in equation (3) by $(I_0)_\theta$, the effective intensity of irradiation for each value of $\theta$, and integration over the whole range of $\theta$ yield the following expression:

$$I = \frac{KM(\nu_0 - \Delta\nu)^4}{\Delta\nu[1 - \exp(-h\Delta\nu/kT)]}$$
$$\times \int (I_0)_\theta \{[45(\bar{\alpha}')^2 + 7\gamma'^2](1 + \cos^2\theta) + 6(\gamma')^2 \sin^2\theta\} \, d\theta \qquad (4)$$

Although it is not practicable to evaluate this integral, because the dependence of $(I_0)_\theta$ upon $\theta$ is unknown, it is possible to factorize the integrand, making use of the equation given in the introductory chapter for the depolarization ratio $\rho$:

$$(\gamma')^2 = \frac{45(\bar{\alpha}')^2}{6 - 7\rho} \qquad (5)$$

Substitution in equation (4) gives

$$I = \frac{KM(\nu_0 - \Delta\nu)^4}{\Delta\nu[1 - \exp(-h\Delta\nu/kT)]} 45(\bar{\alpha}')^2 \int (I_0)_\theta \frac{(6 + 6\cos^2\theta + 6\rho\sin^2\theta) \, d\theta}{6 - 7\rho}$$
$$(6)$$

Determination of the relative intensities of Raman lines having different depolarization ratios calls for irradiation with polarized light. When this is achieved by the Edsall and Wilson[6] method of surrounding the sample tube with a Polaroid cylinder, oriented so that its planes of polarization (defined as the planes of the electric vector) are at right angles to the axis of the tube, the intensity equation can be reduced to the form

$$I = \frac{KM(\nu_0 - \Delta\nu)^4}{\Delta\nu[1 - \exp(-h\Delta\nu/kT)]}[45(\bar{\alpha}')^2 + 7(\gamma')^2] \tag{7}$$

Quantitative determination of true depolarization ratios $\rho$ using this Polaroid cylinder method, together with a low-pressure helical mercury-arc lamp, has been shown to be feasible by Rank and co-workers.[7,8] Finally, combination of equations (7) and (5) yields the following equation for the intensity of the Raman line characterizing the $p$th normal mode of vibration $Q_p$:

$$I_p = \frac{KM(\nu_0 - \Delta\nu)^4}{\Delta\nu[1 - \exp(-h\Delta\nu/kT)]}45\left(\frac{\partial\bar{\alpha}}{\partial Q_p}\right)^2\left(\frac{6}{6 - 7\rho}\right) \tag{8}$$

The foregoing treatment, due largely to Woodward and Long,[5] has thus provided a relationship between the Raman intensity obtained with the most usual experimental arrangement and the quantity $(\partial\bar{\alpha}/\partial Q_p)$. This has perhaps been the most widely used equation for relating intensities to chemical-bond properties such as bond order, degree of covalence, etc., and will be explored further in a later section.

Similar developments of the Placzek[2] polarizability theory by Brandmüller and Moser[9] and by Bernstein and his colleagues[66] have led to the following equation for the Raman intensity for perpendicularly polarized incident light:

$$I_p = \text{const} \cdot Nk_p b_p^2 g_p[45(\bar{\alpha}')^2 + 7(\gamma')^2]I_0 \tag{9}$$

Here $k_p$ signifies $(\nu_0 - \Delta\nu)^4[1 - \exp(-h\Delta\nu/kT)]^{-1}$; $N$ is the number of scattering molecules; $b_p^2 = h/8\pi^2\Delta\nu$ is the square point of the zero point amplitude; and $g_p$ is the degree of degeneracy of the $p$th normal mode. The expression is valid irrespective of the covergence angle of the incident light (i.e., is not dependent on strictly rectilinear irradiation) provided only that the light source has cylindrical symmetry.

An alternative development of equation (4) has been made by Bernstein and Allen,[10] who make the correction for convergent (non-rectilinear) incident radiation by considering the whole incident beam to make some average angle $\bar{\theta}$ with the scattering direction. Then, by use of the expression developed by Martin[11] for the depolarization ratio of a Raman line scattered in a direction making an angle $\bar{\theta}$ with the incident ray, namely,

$$\rho_{\bar{\theta}} = \frac{3(\gamma')^2(1 + \sin^2\bar{\theta}) + [45(\bar{\alpha}')^2 + 4(\gamma')^2]\cos^2\bar{\theta}}{45(\bar{\alpha}')^2 + 7(\gamma')^2} \tag{10}$$

and by substitution in equation (3), the following expression is obtained:

$$I_{\bar{\theta}} = \text{const} \cdot \frac{(\nu_0 - \Delta\nu)^4 N}{\Delta\nu[1 - \exp(-h\Delta\nu/kT)]}[45(\bar{\alpha}')^2 + 7(\gamma')^2](1 + \rho_{\bar{\theta}}) \tag{11}$$

This equation, unlike equations (8) and (9), applies for natural (non-polarized) incident light and leads to an expression for the Raman intensity corrected for convergence of the incident light:

$$\frac{I}{1 + \rho} = \text{const} \cdot \frac{(\nu_0 - \Delta\nu)^4}{\Delta\nu[1 - \exp(-h\Delta\nu/kT)]}[45(\bar{\alpha}')^2 + 7(\gamma')^2] \tag{12}$$

This equation has the same form when $I_{\text{true}}$ and $\rho_{\text{true}}$ are used that it has when $I_{\text{observed}}$ and $\rho_{\text{observed}}$ are used, as has been checked experimentally by Bernstein and Allen.[10] These authors have used equation (12) as a basis for establishing a standard intensity scale which will be discussed later.

The various expressions for Raman intensities developed thus far provide a means for converting measured intensities and depolarization ratios into mean molecular polarizability derivatives $\partial\bar{\alpha}/\partial Q$. Some attempts have been made to relate these derived quantities directly to characteristics such as degree of covalence of bonds involved in the molecular vibration described by the normal coordinate $Q$. More success, however, has been achieved by making use of known molecular symmetry, atomic masses, and bond force constants to break down $\partial\bar{\alpha}/\partial Q$ values into individual bond polarizability derivatives. A simple theory proposed by Wolkenstein[12] treats molecular polarizabilities in terms of contributions resulting from the stretching and change of orientation of the individual bonds of the molecule. The simple theory, developed further by Eliashevich and Wolkenstein,[13] and formulated in matrix terms by Long,[14] involves the basic assumption that the polarizability of a bond is not affected by a change of bond orientation. Determinations of the intensities of vibrations of neopentane by Woodward and co-workers,[15,16] however, indicate that this assumption of the Wolkenstein theory is not entirely valid, and a modification has been introduced to remove this limitation. These points also will be discussed further, but first it should be realized what is implied by the restriction placed on the whole of the foregoing treatment that the incident (exciting) light frequency $v_0$ be well removed from any absorption band of the molecule.

In the introductory chapter, Dr. Woodward has pointed out the distinction between Rayleigh and Raman scattering on the one hand and fluorescence scattering on the other. For the latter phenomenon to occur, the molecule must have a stable vibronic state (characterized by vibrational and electronic quantum numbers) whose energy separation from the ground state of the molecule corresponds exactly to the frequency $v_0$. For a Raman transition between two vibrational states in the same electronic ground state, the intermediate active excited state must be considered. No reference to this intermediate state has been necessary in the theoretical approaches outlined previously, and the role of excited vibronic states in contributing to the molecular polarizability has been obscured. Recent interest in resonance or near-resonance scattering, particularly in the resonance Raman effect, has led to a different approach from that taken by Placzek[2] and subsequent workers. As developed by Albrecht,[17] this alternative approach to the formulation of the Raman effect (and the resonance Raman effect) takes as its starting point the Kramers–Heisenberg dispersion equation and Van Vleck's[18] expansion of this equation to produce fundamental selection rules for Raman spectra. The dispersion theory contains explicit reference to excited vibronic states and, properly formulated, is valid (unlike previously discussed polarizability theory[2]) under resonance conditions. Using the Hertzberg–Teller theory,[19,20] Albrecht has characterized the vibronic interactions among states which can participate in allowed transitions from the ground state. It is shown that these particular interactions are of importance in the dispersion theory of Raman intensities. For resonance or near-resonance conditions, contributions from two neighboring, close-lying, and vibronically mixing virtual electronic states are important. It seems appropriate here to quote the predictions made by the Albrecht theory for the frequency dependence of Raman intensities. If the frequencies characterizing the two virtual electronic states are labeled $v_e$ and $v_s$, respectively, and the frequency of the incident light is $v_0$, the following relationship is obtained with near-resonance conditions:

$$ I \propto v_0^4 \left\{ \frac{v_e v_s + v_0^2}{[(v_e^2 - v_0^2)(v_s^2 - v_0^2)]^2} \right\} \tag{13} $$

When $v_s \rightarrow v_e$ (very close-lying vibronically mixing electronic states), equation (13) reduces to the form

$$I \propto v_0^4 \left[ \frac{v_e^2 + v_0^2}{(v_e^2 - v_0^2)^4} \right] \qquad (14)$$

This equation is identical in form to an equation resulting from the semiclassical theory of Shorygin,[21,22] which has been reviewed by Behringer and Brandmüller[23] and found to agree well with experimental data on Raman intensities excited by frequencies close to electronic absorption bands of molecules. However, when $v_s \gg v_e$, equation (13) resembles more nearly the form

$$I \propto \frac{v_0^4}{(v_e^2 - v_0^2)^2} \qquad (15)$$

which has been found to be relatively less successful[23] than equation (14). When the frequency of the incident light is much lower than the frequencies associated with the virtual transitions, and yet at the same time is much higher than the Raman frequency displacement, then the intensity of the Raman line is simply proportional to $v_0^4$; that is, with $v_e, v_s \gg v_0 \gg \Delta v$, equation (13) reduces to $I \propto v_0^4$. This final form is also obtained from equations (8), (9), and (12), the derivations of which ignored the virtual states, but it is now possible to see the consequences of exciting Raman spectra with light close in frequency to that of allowed electronic transitions within a molecule. In general, one can expect the distribution of intensities in a Raman spectrum to change as resonance is approached, and, moreover, Albrecht's theory enables one to predict which particular vibrational modes of a given symmetry will be enhanced in intensity. This prediction requires a knowledge of the vibronic spectrum of the molecule involved, but, conversely, enhancement of a particular normal mode in the Raman spectrum when approaching resonance can lead to predictions concerning the electronic transition involved. It further follows from the theory that, near resonance, depolarization ratios for totally symmetric vibrations must reach a maximum. Values of depolarization ratios of lines from a number of substituted aromatic nitro compounds having intense absorption bands (allowed vibronic transitions) in the near-ultraviolet region have been measured by Rea.[24] These values, used by Rea as a test of Shorygin's equation, constitute (at least qualitatively) a satisfactory confirmation of this aspect of the Albrecht theory. Further experimental verification of the theory will be discussed in a later section, and further development of the theory is currently in progress.

Chapter 6 should be consulted for an expanded treatment of theories of the resonance Raman effect.

## BOND POLARIZABILITY DERIVATIVES

The intensity theories developed in the previous section enable values of the polarizability derivatives $\bar{\alpha}'$ and $\gamma'$ to be calculated from experimental values of Raman line intensities and depolarization factors. However, these are derivatives with respect to the normal modes of vibration of the molecule and, as such, can provide information directly concerning particular bonds in the molecule only if the normal mode concerned involves only that particular type of bond. Removal of this restriction necessitates the breakdown of the polarizability and anisotropy derivatives into components characteristic of particular bonds in the molecule. This type of breakdown requires detailed analysis of the normal modes in terms of contributions from motions of individual atoms within the molecule. In turn, this requires knowledge of the molecular symmetry, atomic masses, and interatomic forces. The well-known Wilson $FG$ matrix technique for analysis of normal modes of vibration[25,26] has been applied to this bond polarizability problem by Long,[14] making use of the earlier developments due to Wolkenstein[12] and Eliashevich.[13] The basis of this treatment is the assignment of a characteristic bond polarizability to each bond in the molecule. This bond polarizability is then defined completely in terms of a polarizability ellipsoid having its major axis coincident with the bond direction. Three polarizability components $\alpha_{uv}$ are assigned to correspond with the principal axes of the ellipsoid, $u$ referring to the $u$th bond and $v$ having the values 1, 2, or 3, depending on whether the component is along the bond or at right angles to the bond. The assumptions are made that the bond direction represents the electrical axis of the bond under both equilibrium and nonequilibrium conditions, that individual bond polarizabilities are independent of environment, and that these individual bond polarizabilities are additive. This last assumption gives an expression for the components of the overall mean molecular polarizability (referred to a space-fixed Cartesian coordinate system $X, Y, Z$, as previously):

$$\alpha_{XY} = \sum_{u=1}^{U} \sum_{v=1}^{3} [\alpha_{uv}(uv\ X)(uv\ Y)] \tag{16}$$

The first sum is taken over all $U$ bonds in the molecule; $(uvX)$ is the direction cosine of the $uv$-axis of the polarizability ellipsoid of the $u$th bond with the positive $X$-axis, and ($uvY$) has similar meaning. When a bond is cylindrically symmetrical about its own direction axis, $\alpha_{u2} = \alpha_{u3}$. With application of the normalization and orthogonality conditions to the direction cosines and performance of the sum over $v$, for a bond with cylindrical symmetry, equation (16) becomes

$$\alpha_{XY} = \sum_{u=1}^{U} [(\alpha_{u1} - \alpha_{u2})(uX)(uY) + \alpha_{u2}\delta_{XY}] \qquad (17)$$

where $\delta_{XY}$ is the Kronecker delta (equal to 1 if $X = Y$, equal to 0 if $X \neq Y$).

The problem may be further simplified at this stage by assuming that stretching one bond in a molecule has no effect on the polarizability components of other bonds and that the polarizability components of a bond are unaffected by changes in the orientation of the bond. These conditions may be expressed mathematically as

$$\frac{\partial \alpha_{uv}}{\partial r_u} = \frac{\partial \alpha_{uv}}{\partial r_{u'}} \delta uu' \qquad (18)$$

and

$$\frac{\partial \alpha_{uv}}{\partial(u'X)} = 0 \qquad (19)$$

for $u = u'$ and $u \neq u'$. Here $r_u$ represents the length of the $u$th bond. These approximations are drastic, corresponding roughly to the valence-force approximation often used in vibrational analysis.[26] Like the latter, the approximations allow a relatively simple and often useful treatment of real molecules, although cases will be found where experimental data demand a more general treatment.

The expression of the normal coordinates in terms of Cartesian displacements of the $N$ atoms in the molecule may be summarized as follows: Using the standard matrix notation,[26] the normal coordinates are related to internal coordinates (the $3N - 6$ changes of bond lengths, interbond angles, etc., represented by a column matrix $R$):

$$R = LQ \qquad (20)$$

In turn, this internal coordinate set $R$ may be related to the $3N$ Cartesian displacements $x$:

$$x = AR \qquad (21)$$

where $A$ is a $(3N - 6)$ by $(3N)$ matrix. These Cartesian displacements

$x$ are useful for the definition of the tensor quantities $\alpha_{XY}$, etc., which are referred to space-fixed axes and so cannot be defined in terms of internal coordinates. Without neglecting possible rotation–vibration interactions,[14] the reverse of the transformation in equation (21) may be written as

$$R = Bx \tag{22}$$

where $B$ is a $(3N)$ by $(3N - 6)$ matrix, the entries in which may be determined as follows: Two bonds $u$ and $u'$ are defined, the former joining atoms $i$ and $j$, with $i$ otherwise nonbonded, and the latter joining atoms $j$ and $i'$, with $i'$ otherwise nonbonded. The angle between these bonds is $\theta_{uu'}$. Then entries in the $B_r$ part, dealing only with changes in bond length, have the form

$$\left|\frac{\partial r_{ij}}{\partial x_i}\right| = |uX| \tag{23}$$

with

$$\frac{\partial r_{ij}}{\partial x_i} = -\frac{\partial r_{ij}}{\partial x_j} \tag{24}$$

the signs being determined by inspection. Equation (24) applies to entries for atoms $i$ and $j$, or $j$ and $i'$. For the $B_\theta$ part of $B$, dealing only with changes of interbond angle $\theta$,

$$\frac{\partial \theta_{uu'}}{\partial x_i} = \frac{1}{\sin \theta_{uu'}} \sum_{XYZ} \left[(u'X)\frac{\partial(uX)}{\partial x_i}\right] \tag{25}$$

$$\frac{\partial \theta_{uu'}}{\partial x_{i'}} = -\frac{1}{\sin \theta_{uu'}} \sum_{XYZ} \left[(uX)\frac{\partial(u'X)}{\partial x_{i'}}\right] \tag{26}$$

and

$$\frac{\partial \theta_{uu'}}{\partial x_j} = -\frac{1}{\sin \theta_{uu'}} \sum_{XYZ} \left[(uX)\frac{\partial(u'X)}{\partial x_j} + (u'X)\frac{\partial(uX)}{\partial x_j}\right] \tag{27}$$

Since matrices $A$ and $B$ are rectangular, one is not the inverse of the other, but they are related through an identity matrix of order $(3N - 6)$, and thus

$$AB = E_{3N-6}$$

In addition, Crawford and Fletcher[27] have established the further relationship between $A$ and $B$ as follows:

$$A = M^{-1}\tilde{B}G^{-1} \tag{28}$$

where $M^{-1}$ is a diagonal $(3N)$ by $(3N)$ matrix of the reciprocal atomic masses, $\tilde{B}$ is the transpose of $B$, and $G^{-1}$ is the inverse of the Wilson $G$ matrix formed from the internal coordinates.

From equations (20), (21), and (28), a relationship between the Cartesian displacements $x$ and the normal coordinates $Q$ emerges:

$$x = M^{-1}\tilde{B}G^{-1}LQ \tag{29}$$

which, using the property $L\tilde{L}= G$, may be reduced to

$$x = M^{-1}\tilde{B}\,\tilde{L}^{-1}Q \tag{30}$$

It is now possible to express the dependence of the polarizability component $\alpha_{XY}$ on the Cartesian displacements, using a intermediate set of intensity coordinates, represented by the column matrix $I$, such that

$$\alpha_{XY} = J_{XY}I \tag{31}$$

where the entries in the row matrix $J_{XY}$ are of the following type:

$$\frac{\partial \alpha_{XY}}{\partial r_u} = \left(\frac{\partial \alpha_{u1}}{\partial r_u} - \frac{\partial \alpha_{u2}}{\partial r_u}\right)(uX)(uY) + \frac{\partial \alpha_{u2}}{\partial r_u}\delta_{XY} \tag{32}$$

and

$$\frac{\partial \alpha_{XY}}{\partial(uX)} = (\alpha_{u1} - \alpha_{u2})(uY) \tag{33}$$

The relationship between these intensity coordinates and the Cartesian displacements takes the form

$$I = Kx \tag{34}$$

The $K$ matrix may be partitioned into submatrices dealing separately with bond stretches $K_r$ and changes in direction cosines $K_\phi$. Using the same model as for the $B$ matrix entries, the elements of $K_\phi$ take the form

$$\left|\frac{\partial(uX)}{\partial x_i}\right| = \left|\frac{1 - (uX)^2}{r_u}\right| \tag{35}$$

and

$$\left|\frac{\partial(uX)}{\partial y_i}\right| = \left|\frac{(uX)(uY)}{r_u}\right| \tag{36}$$

where the signs are determined by inspection, and entries for atoms

$i$ and $j$, or $j$ and $i'$ are related by the expression

$$\frac{\partial(uX)}{\partial x_i} = -\frac{\partial(uX)}{\partial x_j} \tag{37}$$

Now, combining equations (30), (31), and (34), the polarizability component $\alpha_{XY}$ may be written

$$\alpha_{XY} = J_{XY}KM^{-1}\tilde{B}\,\tilde{L}^{-1}Q \tag{38}$$

which reduces to

$$\alpha_{XY} = J_{XY}KALQ \tag{39}$$

The matrices $J$, $K$, $A$, and $L$ may all be partitioned into submatrices, as already shown for $B$ and $K$, and equation (39) may be rewritten as

$$\alpha_{XY} = [J_{XY,r} | J_{XY,\phi}]\left[\begin{array}{c}K_r\\\hline K_\phi\end{array}\right][A_r, A_\theta]\left[\begin{array}{c}L_r\\\hline L_\theta\end{array}\right][Q] \tag{40}$$

Recognizing that $K_r A_r = E_U$, the $K_r$ submatrix is eliminated from equation (40) multiplied out to give

$$\alpha_{XY} = J_{XY,r}L_r Q + J_{XY,\phi}K_\phi ALQ \tag{41}$$

At this point in the analysis it is convenient to make use of molecular symmetry and to distinguish between polarizability derivatives for totally symmetric ($A_1$) modes of vibration and nontotally symmetric modes. For this purpose, the internal coordinate set $R$ is divided into sets of equivalent coordinates, designating $q_i\delta$ as the $\delta$th coordinate of the $i$th equivalent kind, with $\delta = 1, 2, \ldots, \mu_i$. The first $j$ coordinates are designated as bond stretches, the remainder being changes in bond angles. The relation between the internal coordinates and the normal coordinates, equation (20), may now be written as

$$q_{i\delta} = \sum_p \sum_\sigma l_{i\delta,p\sigma}Q_{p\sigma}^{(k)} \tag{42}$$

where $Q_{p\sigma}^{(k)}$ is the $\sigma$th normal mode of vibration of the symmetry species (irreducible representation) $k$, giving rise to the vibrational frequency $p$. The $A_1$ modes are distinguished by coefficients $l_{i\delta,p\sigma}^{(A_1)}$ which are independent of $\delta$, so that from the orthogonality condition on modes belonging to different symmetry species,

$$\sum_i \sum_\delta l_{i\delta,p\sigma}^{(A_1)} l_{i\delta,p'\sigma'}^{(k)} = 0 \qquad \text{for } k \neq A_1 \tag{43}$$

it is seen that

$$\sum_{\delta} l_{i\delta, p\sigma}^{(k)} = 0 \qquad \text{when } k \neq A_1 \qquad (44)$$

Then, from equation (41) the following expression is obtained:

$$\left(\frac{\partial \alpha_{XY}}{\partial Q_{p\sigma}}\right) = \sum_{i} \sum_{\substack{\delta \\ i \leq j}} \frac{\partial \alpha_{XY}}{\partial q_{i\delta}} l_{i\delta, p\sigma} + \sum_{\phi} \sum_{\theta} \sum_{i} \sum_{\delta} \frac{\partial \alpha_{XY}}{\partial I_{\phi}} \frac{\partial I_{\phi}}{\partial x_{\theta}} \frac{\partial x_{\theta}}{\partial q_{i\delta}} l_{i\delta, p\sigma} \qquad (45)$$

The expression for the mean molecular polarizability,

$$\bar{\alpha} = \tfrac{1}{3}(\alpha_{XX} + \alpha_{YY} + \alpha_{ZZ}),$$

with the components written in the form of equation (17), may be differentiated with respect to a normal mode $Q$, to give

$$\bar{\alpha}' = \tfrac{1}{3} \sum_{u} \{(\alpha'_{u1} - \alpha'_{u2})[(uX)^2 + (uY)^2 + (uZ)^2] + 3\alpha'_{u2}\} \qquad (46)$$

The contribution to this $\bar{\alpha}'$ from the second summation in equation (45) is always zero, whatever the symmetry of the normal mode, since the direction cosines involved obey normalization and orthogonality conditions. From these considerations, the following expressions for the mean molecular polarizability derivatives are obtained:

$$(\bar{\alpha}_{p\sigma})^{(A_1)} = \tfrac{1}{3} \sum_{i}^{i=j} (\alpha'_{i1} + 2\alpha'_{i2}) \mu_i l_{i\delta, p\sigma}^{(A_1)} \qquad \text{for } k = A_1 \qquad (47)$$

and

$$(\bar{\alpha}'_{p\sigma})^{(k)} = 0 \qquad \text{for } k \neq A_1 \qquad (48)$$

It follows from equation (48) that for all modes other than the $A_1$ type, the degree of depolarization $\rho$ must equal $\tfrac{6}{7}$.

Finally, since this treatment is for bonds of cylindrical symmetry, with $\alpha_{u2} = \alpha_{u3}$, any bond polarizability derivative may be expressed as follows:

$$\alpha'_u = \left(\frac{\partial \alpha_{u1}}{\partial r_u} + 2\frac{\partial \alpha_{u2}}{\partial r_u}\right) \qquad (49)$$

and the bond anisotropy written as

$$\gamma_u = (\alpha_{u1} - \alpha_{u2}) \qquad (50)$$

from which the derivative is simply obtained:

$$\gamma'_u = \left(\frac{\partial \alpha_{u1}}{\partial r_u} - \frac{\partial \alpha_{u2}}{\partial r_u}\right) \qquad (51)$$

The application of this simple bond-polarizability theory usually involves the use of symmetry coordinates $S$ obtained from the transformation

$$S = UR \qquad (52)$$

as linear combinations of the internal coordinates. Taking full advantage of molecular symmetry, these coordinates form a basis for a completely reduced representation of the molecular point group, and provide the simplest form of the secular equation. The transformation matrix $U$ is orthogonal, i.e., $U\tilde{U} = E$. The relationship between normal coordinates and symmetry coordinates is seen from equations (20) and (52) to be

$$S = \mathscr{L}Q \qquad (53)$$

where

$$\mathscr{L} = UL \qquad (54)$$

The outline of the bond-polarizability theory provided here will serve as a basis for much of the discussion of results in the following section. However, in anticipation of some of the problems which will become apparent from a consideration of experimental data, it is worth noting here a part of the theory which has proven inadequate. This is the assumption that the polarizability of a bond is not affected by a change of bond orientation. Careful work by Woodward and colleagues[16,28] on the determination of intensities of the totally symmetric vibrations of gaseous neopentane, $C(CH_3)_4$, and fully deuterated neopentane, $C(CD_3)_4$, has established the need for a modification in which the above assumption is abandoned. The critical test suggested by Chantry and Woodward[29] is whether a force field can be found which will account for both the observed intensities (on the basis of the simple Wolkenstein theory) and the observed frequencies. Taylor and Woodward have shown that such a field cannot be found for $C(CH_3)_4$ and $C(CD_3)_4$, although the introduction of an extra parameter that takes into account changes in polarizability with bond orientation enabled these authors to account for observed intensities and frequencies completely satisfactorily. This modified Wolkenstein theory has also been applied to observed intensities from $CHCl_3$ and $CDCl_3$.[30] However, in this case, when the uncertainties usually present in the normal coordinate transformation matrix $L$ are eliminated by the isotopic substitution data, the simple and modified bond-polarization theories are found to be in good agreement.

In conclusion, it seems reasonable to say that the state of the theory is such that, for simple molecules at least, it gives values of bond-polarizability derivatives which are transferable from one molecule to another. An analogy with characteristic "group frequencies" can be drawn. In favorable cases, intensity data from one molecule can be used to predict intensities from another. However, it must be recognized that there are certain to be many limitations, and great care must be taken in interpretation of bond-polarizability derivatives in terms of the distribution of electrons in a chemical bond. The simple picture can, in general, be expected to be only approximately correct, with such effects as changes in lone-pair hybridization, bond–bond interactions, and nonbonded atom perturbations probably invalidating conclusions based on small changes in bond-polarizability derivatives from one molecule to another. For double bonds such as that in ethylene, a further complication arises as a result of the non-cylindrical symmetry of the $C{=}C$ bond. In order to see what can be done, it is necessary now to examine some experimental results in the light of the theories developed.

## BOND TYPES FROM RAMAN INTENSITIES—FROM COVALENT MOLECULES TO ION PAIRS

Considerable experimental difficulties are associated with the determination of Raman intensities, particularly if the intensities are to be placed on an absolute scale. Some indication of the precautions needed will be given in the following discussion of the work of various authors whose results are presented.

Woodward and Long's[5] work on the intensities of the totally symmetric stretching modes of Group IV tetrahalides provides a good starting point for this discussion. Photographic plates on which their spectra were recorded were calibrated by means of a tungsten-band lamp operated under standard conditions, and microphotometer traces were made of the Raman lines whose relative intensities were to be measured. The measurement of absolute Raman intensities always requires the use of a standard whose molar intensity is known or can be calculated, and Woodward and Long used carbon tetra-chloride as an internal standard for their work. That is, the liquids investigated all contained a known amount of $CCl_4$, and all line intensities (measured as products of peak height and width at half-peak height) then were measured relative to the $CCl_4$ $v_1(A_1)$ line. The possibility that interactions between the components of the mixtures (solvent effects) might bring about changes in molar intensities was eliminated by studying at least two different molar ratios for each mixture and finding good agreement between the $(I_2/I_1)(M_1/M_2)$ values in each case. The use of an internal standard, demonstrated by this work, is generally preferable to the method of determining relative intensities from successive experiments with the separate single substances. Reproducing exactly the same conditions of sample alignment, lamp intensity, exposure time, etc., is virtually impossible, but these problems are eliminated with the internal-standard technique. Of course, this internal-standard method will fail when there are strong intermolecular interactions between the mixture components, and this point will be pursued further subsequently.

The results found for the corrected molar intensities relative to that of the $v_1(A_1)$ line of $CCl_4$ taken as unity are as follows: $SiCl_4$, 1.0(4); $GeCl_4$, 3.2; $SnCl_4$, 4.8; $TiCl_4$, 7.5; $CBr_4$, 3.8. Previous work by Welsh, Crawford, and Scott[31] involving intensity determinations from calibrated photographic plates using similar precautions and corrections produced intensity values (relative to $CCl_4$) for $SnCl_4$ and $SnBr_4$ of 5.0 and 14, respectively. Since all of these figures are good to no better than $\pm 5\%$, the agreement for the common member, $SnCl_4$, is seen to be satisfactory. Both sets of results show that the molar scattering is greater when an atom in the molecule is replaced by a heavier atom of the same chemical group, although the transition-metal halide, $TiCl_4$, is seen to be an exception to this general observation.

Conversion of relative intensities to relative polarizability derivatives $\bar{\alpha}'$ is particularly simple in the case of the $A_1$ modes of the Group VI tetrahalides, since for these, $\gamma' = 0$, and reduction of equation (4) yields the relationship

$$\frac{I_2}{I_1} = \frac{M_2}{M_1}\left(\frac{v_0 - \Delta v_2}{v_0 - \Delta v_1}\right)^4 \frac{\Delta v_1}{\Delta v_2} \frac{1 - \exp(-h\Delta v_1/kT)}{1 - \exp(-h\Delta v_2/kT)}\left(\frac{\bar{\alpha}'_2}{\bar{\alpha}'_1}\right)^2 \qquad (55)$$

For the $A_1$ lines compared, all depolarization ratios were assumed by Woodward and Long to be equal to zero, so that no corrections to their measured intensities were necessary on this account. The $\bar{\alpha}'$ values obtained are as follows; $CCl_4$, 1.00 (standard); $SiCl_4$, 0.96; $GeCl_4$, 1.6; $SnCl_4$, 1.9; $TiCl_4$, 2.4; $CBr_4$, 2.0; $SnBr_4$, 3.3. These results are summarized in Table I.

## TABLE I

### Molecular Polarizability Derivatives for the Totally Symmetric Stretching Modes of Group IV Halides

| Halide | Derivatives | | | |
|--------|-------------|-----|------------------|----------------------------|
|        | $\nu_1(A_1)$, cm$^{-1}$ | $\bar{\alpha}'$ | $\bar{\alpha}'_{emp}$ | Percentage covalence, $p$ |
| $CCl_4$  | 459 | (1.00) | (1.00) | 94   |
| $SiCl_4$ | 422 | 0.96   | 1.0    | 70   |
| $GeCl_4$ | 396 | 1.6    | 1.5    | 65.5 |
| $SnCl_4$ | 367 | 1.9    | 2.0    | 65.5 |
| $CBr_4$  | 265 | 2.0    | 1.9    | 98   |
| $SnBr_4$ | 220 | 3.3    | 2.9    | 74   |

The values of $\bar{\alpha}'$ found were shown by Woodward and Long to be at variance with the earlier assumption of Hansen-Damaschun[32] that $\bar{\alpha}'$ values are simply proportional to the covalent character of the bonds involved. This assumption would require, for example, that $SnCl_4$ be a more covalent molecule than $CCl_4$, since $SnCl_4$ gives the higher $\bar{\alpha}'$-value. However, the observed $\bar{\alpha}'$-values were found to agree quite well with values calculated from the empirical equation

$$\bar{\alpha}'_{emp} = Cp(Z_X + Z_Y) \tag{56}$$

where $\bar{\alpha}'_{emp}$ is the empirical value of $\bar{\alpha}'$, $Z_X$ and $Z_Y$ are the atomic numbers of the atoms in the molecule $XY_4$, $p$ is the percentage covalent character of the bond, and $C$ is a proportionality constant determined by the $CCl_4$ standard. Values of $p$ were calculated according to the method proposed by Pauling,[33] namely,

$$p = 100 \cdot \exp[-\tfrac{1}{4}(x_X - x_Y)^2] \tag{57}$$

where $x_X$ and $x_Y$ are Pauling electronegativity values. The numbers calculated for $p$ and $\bar{\alpha}'_{emp}$ are also listed in Table I. Woodward and Long do not suggest any satisfactory theoretical interpretation of their empirical rule, and they point out that $TiCl_4$ does not conform ($\bar{\alpha}' = 2.4$, $\bar{\alpha}'_{emp} = 1.1$). However, these results demonstrate rather clearly that there is a correlation between Raman intensities and the nature of the chemical bonding involved, and a simple dependence on degree of covalence and sum of atomic numbers (perhaps to be related to atomic polarizabilities) is encouraging at this stage.

As with previous theories,[32-35] equation (56) leads to the conclusion that a molecule with purely ionic bonds would give Raman lines of zero intensity, i.e., no Raman lines. This implies that the molecular polarizability in such a case will be independent of the interionic distances, an implication which cannot be supported theoretically, since intramolecular mutual polarization effects in an assembly of ions will certainly depend on the interionic distances. Hence, even for a completely ionic case, the derived polarizability should be finite and a vibrational Raman effect should exist.

This latter point takes on a particular significance when applied to the consideration of Raman spectra of ion pairs known to exist in many electrolyte solutions. It is evident that one might hope to distinguish direct ion pairs (ions in intimate contact, though with little or no electron sharing) from solvent-separated ion pairs (ions held together by electrostatic forces transmitted through one or more intervening solvent molecules) through their Raman spectra. However, Raman line intensities characterizing ionic bonds are invariably weak, and this distinction has, in fact, proven extremely difficult to achieve.

Raman evidence for ion-pair formation has usually been somewhat less direct than the observation of a line having its origin in the vibration of the interionic bond. Polyatomic ions have commonly been used as indicators for ion pairing. The distortion of the normal symmetry of a polyatomic ion, due to a closely neighboring ion, gives rise to a new set of Raman lines characterizing the distorted (ion-paired) ion. For example, such distortion of the nitrate ion in aqueous solution by a series of metal ions has been reported by a number of authors,[36-39] the splitting of the two $E$-type modes (in free $NO_3^-$, of $D_{3h}$ symmetry) into pairs of $A$- and $B$-type modes (in bound $NO_3^-$, of $C_{2v}$ symmetry) serving to indicate pairing of the $NO_3^-$ and metal ions in solution. However, in most of the studies made, no lines have been detected which could be assigned to anion–cation stretching modes. Even in molten electrolyte systems there seems to be little evidence for Raman lines characterizing interionic bonds, although many examples of distortion of polyatomic anion symmetry in melts have been reported.[40] Infrared absorption bands observed[41] in the 200–350 cm$^{-1}$ region from several alkali metal nitrates in the molten state are presumably the result of strong anion–cation interactions between the nitrate ion and its nearest neighbors, but these have been described as "lattice-like" vibrations and have not been detected in the Raman spectra of such melts.

A theoretical treatment of the problem of Raman intensities from electrostatically bound ion pairs given by George, Rolfe, and Woodward[42] has indicated that a factor as high as 100 probably separates intensities from vibrations of electrostatic and covalent linkages. In an attempt to find a vibrational Raman effect from an ion pair, these authors studied concentrated thallous hydroxide solutions, but were unable to detect a Tl—OH stretching frequency, even though the dissociation constant for the $Tl^+OH^-$ species is known to be as low as 0.15,[43] so that their solutions contained a $0.9M$ concentration of the ion pairs. Solubility work on this thallous hydroxide system[43] has shown beyond reasonable doubt that the associated species in these solutions is an intimate $Tl^+OH^-$ ion pair (no water molecules separate the ions), so that the Raman results strongly suggest that the appearance of a Raman line at all requires some degree of electron sharing or covalent character in the bond characterized. This view is further supported by the absence of a Tl—O line in the spectrum of aqueous solutions of dimethyl thallic hydroxide,[44] where evidence for intimate ion pairing is also present in the frequency shifts of lines characterizing the $(CH_3)_2Tl^+$ ion. However, Raman spectra from analogous solutions of methyl mercuric hydroxide do show a polarized line at $511\ cm^{-1}$, which finds its origin in the Hg—O stretching vibration.[44] Accordingly, this may be taken as evidence of covalent Hg—O bonding. No estimation of the percentage covalent character of the Hg—O bond has been made from the intensity of the $511\ cm^{-1}$ line, but it does appear that a meaningful distinction can now be made between electrostatic and covalent "ion pairs." Thus, the recent statement by Lee and Wilmshurst,[45] "The distinction between an ion pair with ions in mutual contact and a molecular species is probably one of semantics," would appear to be unjustified. Lee and Wilmshurst, from the results of their Raman studies of aqueous solutions of a series of monovalent metal nitrates, carbonates, sulfates, and perchlorates, conclude that ion pairs separated by a single water molecule are probably present in many concentrated aqueous electrolyte solutions. Their conclusion also provides a reasonable explanation of Hester and Plane's[38] findings that most metal sulfates, although known to be highly associated in aqueous solution,[46,47] show little or no distortion of the tetrahedral $SO_4^{-2}$ symmetry.

Further Raman evidence for covalent bonding in species resulting from interaction of ions in aqueous solutions is provided by Goggin

## TABLE II
### Raman Lines Characterizing Covalently Bound Metal–Ion Hydration Spheres[50]

| Salt | Raman line, $cm^{-1}$ |
|------|----------------------|
| $Cu(NO_3)_2$ | 440 |
| $Zn(NO_3)_2$ | 390 |
| $Hg(NO_3)_2$ | 380 |
| $Mg(NO_3)_2$ | 370 |
| $In(NO_3)_3$ | 410 and 460 overlapping bands |
| $CuSO_4$ | 440 |
| $MgSO_4$ | 360 |
| $ZnSO_4$ | 400 |
| $Ga_2(SO_4)_3$ | 475, shoulder on $SO_4^{-2}$ line |
| $In_2(SO_4)_3$ | 350–550 |
| $Tl_2SO_4$ | 470 |
| $Cu(ClO_4)_2$ | 440 |
| $Hg(ClO_4)_2$ | 380 |
| $In(ClO_4)_3$ | $\sim$420, shoulder on $ClO_4^-$ line |
| $Mg(ClO_4)_2$ | 360 |

and Woodward's[48] work on methyl mercuric perchlorate and nitrate. These authors were able to assign a line at $464\,cm^{-1}$ to the Hg—O stretching mode in the aqueous solution species $CH_3Hg—OH_2^+$, and a line at $292\,cm^{-1}$ to a similar Hg—O mode in $CH_3Hg—ONO_2$. Both the lines were polarized. The latter nitrate species was found to be in equilibrium with free $NO_3^-$ and the hydrated cation $CH_3HgOH_2^+$ in aqueous solution, but no sign of perchlorate association was found.

The semiorganic nature of the cation used in the above work may well influence the nature of the bonding in the associated species. It is interesting, therefore, to find similar Raman evidence for covalent bonding of water and anions to a simple metal ion. A complete vibrational analysis of Raman spectra from aqueous solutions of indic nitrate, sulfate, and perchlorate has revealed such bonding.[49] Polarized Raman lines due to In—O stretching modes occur at $270\,cm^{-1}$ and $255\,cm^{-1}$, the former caused by In—$ONO_2$ and the latter by In—$OSO_3$ vibrations. Aquo-indium Raman lines were also found in each of these indic solutions, showing that electron sharing is important in these aquo complexes. An ion–dipole interaction is clearly an inadequate description of the bonding of water molecules to $In^{+3}$. Moreover, it appears that many other multiply charged cations are similarly covalently hydrated in aqueous solution. Hester and Plane's[50] observation of low-frequency polarized Raman lines from a wide variety of types of salt solution suggests that some degree of electron sharing in hydrated metal ion $M—OH_2$ bonds is quite common. Table II lists the frequencies of some (metal–water) lines observed from nitrate, sulfate, and perchlorate salts of various metals after the lines due to the oxyanions[51,52] and to the solvent water[53] have been subtracted. The assignment of these lines to metal–water modes is supported by work[54,55] on solid hydrates containing the ions $Mg(H_2O)_6^{+2}$ and $Zn(H_2O)_6^{+2}$. The finite intensities of the lines are evidence of covalent bonding of the hydration spheres.

## BOND ORDERS

From the preliminary survey of bond types contained in the previous section, it is clear that in favorable cases Raman line intensities can be related to the degree of covalent character present in a bond. It will be demonstrated in this section that there is considerable evidence that intensities can give further information on bond orders.

Yoshino and Bernstein,[56] from a series of intensity measurements on gaseous hydrocarbons, came to the interesting conclusion that bond-polarizability derivatives $(\partial\bar{\alpha}/\partial r)$ have values which are simply proportional to the number of bonding electrons in CH and CC bonds, i.e., to the bond multiplicity or bond order for CC bonds. It will be worthwhile to examine this work in some detail.

Great care was taken by Yoshino and Bernstein to ensure that their measured Raman intensities were properly corrected before being used to calculate polarizability derivatives. Measurements of integrated band intensities were made from spectra recorded photoelectrically.[57] In order to get sufficient intensity, a multiple-reflection sample cell that contained the hydrocarbon gases at pressures of 1–2 atm was used and was irradiated by six watercooled Toronto-type[58] mercury-arc lamps. Sheets of Polaroid film were used according to the method of Edsall and Wilson[6] for obtaining depolarization ratios. Prior to measuring a band intensity, the background spectrum measured for the evacuated cell under the conditions used for obtaining the Raman spectrum was subtracted from the gas spectrum. Then the measured (observed) values of intensities $I_{obs}$ and depolarization ratios $\rho_{obs}$ were corrected for the following instrumental factors:

1. A convergence error, due to the incident light not being strictly at right angles to the direction of observation.
2. Lack of equal intensity of illumination on all sides of the sample tube.
3. Different transmission characteristics of the spectrograph for light polarized parallel to and perpendicular to the direction of the length of the slit.
4. Different photomultiplier detector sensitives to light polarized as in (3).
5. Polarization of the scattered light by reflections at the mirrors in the multiple-reflection cell, due to the nonrectilinearity of the incidence on the mirror surfaces.
6. Lack of complete polarization of the incident light, due to Polaroid inefficiency and reflection and refraction by the glass Raman cell surface after the light passed through the Polaroid.
7. Variation with wavelength of $S(\lambda)$, the product of the sensitivity of the photomultiplier and transmission of the spectrometer to light of a definite polarization.
8. Variation with wavelength of $M(\lambda)$, the intensity enhancement produced by the multiple-reflection cell.

All of these corrections were determined experimentally by Yoshino and Bernstein,[57] who showed that their method for obtaining true intensities from those measured was not in error by more than a few percent. This was shown through a comparison of their measured values (after correction) with theoretical values of intensity ratios of hydrogen rotational lines,[59] the $H_2$ and $D_2$ vibrational lines, and the depolarization ratio of the symmetrical CC stretching band of neopentane.

In order to place their Raman intensity values on an absolute scale, Yoshino and Bernstein used the method of comparison with Rayleigh line intensities. A $NiCl_2$ filter solution of predetermined transmission cut down the Rayleigh intensities at 4358 Å from neopentane, $n$-butane, $n$-propane, ethane, methane, and hydrogen, so that they could be compared with the intensity of the methane $\nu_1(A_1)$ Raman line. The square roots of the observed Rayleigh intensities plotted against the polarizabilities of the molecules used gave a straight line, in agreement with the theoretical prediction that $I \propto \bar{\alpha}^2$ for Rayleigh scattering by gases. The value obtained for the intensity ratio of the methane $\nu_1$ band to the Rayleigh line from neopentane was $1.37 \times 10^{-4}$, from which the following relation was derived:

$$\frac{I_{\text{Raman}_A}}{I_{\text{Rayleigh}_B}} = 1.37 \times 10^{-4} \times \frac{I_{\text{Raman line relative to methane } \nu_1}}{(n-1)^2_{\text{B gas}}/(n-1)^2_{\text{neopentane gas}}}$$

where the term $I_{\text{Raman}_A}$ is the intensity of a Raman band of compound A, $I_{\text{Rayleigh}_B}$ is the intensity of a Rayleigh line of compound B, and $n$ represents the refractive index.

For a totally symmetrical vibration (for which $\gamma' = 0$), the ratio of the Raman band intensity to that of a Rayleigh line takes on the simple form[2]

$$\frac{I}{I_0} = \left(\frac{v_0 - \Delta v}{v_0}\right)^4 \left(\frac{\bar{\alpha}'}{\bar{\alpha}}\right)^2 \frac{h}{8\pi^2 \Delta v[1 - \exp(-h\Delta v/kT)]} \tag{58}$$

where notation is the same as that of the first section of this chapter. Using the value of $1.37 \times 10^{-4}$ for the intensity ratio of the methane $v_1$ band to the Rayleigh line from neopentane in equation (58), Yoshino and Bernstein determined for the methane $v_1$ band the value of $\bar{\alpha}' = 2.08 \times N^{\frac{1}{2}} \times 10^{-16}$ cm$^2$/g$^{\frac{1}{2}}$, where $N$ is Avogadro's number. The intensities of the bands due to totally symmetric C—H and C—C vibrations in ethane, neopentane, benzene, ethylene, and acetylene were then measured relative to the $v_1$ band of methane in order to obtain the $\bar{\alpha}'$ values listed in Table III. Unlike the $\bar{\alpha}'$ values listed in Table I, the values given in Table III cannot be compared directly because the normal coordinates involved are different in each case. These values must be transformed into bond-polarizability derivatives $\partial\bar{\alpha}/\partial r$ for meaningful direct comparison. Yoshino and Bernstein[56] used the methods outlined in the second section of this chapter,

## TABLE III
### Absolute Magnitudes of
$\bar{\alpha}' \times N^{-\frac{1}{2}} \times 10^{16}$ (cm$^2$/g$^{\frac{1}{2}}$)     for
### Some Hydrocarbon Molecules[56]

| Molecule | C—H | C—C |
|----------|------|------|
| Methane | 2.08 | — |
| Ethane | 2.65 | 0.41 |
| Neopentane | 3.61 | 0.65 |
| Benzene | 2.31 | 1.20 |
| Ethylene | 1.96 | 0.72 |
| Acetylene | 0.90 | 1.36 |

**TABLE IV**

Bond-Polarizability   Derivatives   $\partial\bar{\alpha}/\partial r$ (Å$^2$)   for
Some Hydrocarbon Molecules

| Molecule | C—H | | C—C | |
|---|---|---|---|---|
| | Ref. 56 | Ref. 66 | Ref. 56 | Ref. 66 |
| Methane | 1.04 | 1.03 | | |
| Ethane I | 1.08 | | 0.92 | |
| Ethane II | 1.10 | | 1.37 | |
| Neopentane | 1.06 | | 1.17 | |
| Benzene | 1.00 | 1.16 | 1.54 | 1.88 |
| Ethylene | 1.04 | | 1.89 | |
| Acetylene | 1.02 | 1.05 | 2.94 | 3.36 |

taking values of force constants from the literature[60-65] to calculate
the entries in the $L$ matrices used in the transformation of symmetry
coordinates to normal coordinates. These enabled the quantities
$\partial\bar{\alpha}/\partial S$, the mean molecular polarizability derivatives with respect to
symmetry-coordinates, to be determined from the $\bar{\alpha}'$-values through the
relationship

$$\left(\frac{\partial\bar{\alpha}}{\partial S_t}\right) = \sum_k L_{kt}^{-1}\left(\frac{\partial\bar{\alpha}}{\partial Q_k}\right) \tag{59}$$

where $Q_k$ refers to the $k$th normal coordinate, and $S_t$ to the $t$th symmetry coordinate. For the CH and CC stretching modes considered,
transformation of $\partial\bar{\alpha}/\partial S$ values into bond polarizability derivatives
$\partial\bar{\alpha}/\partial r$ takes the simple form

$$\left(\frac{\partial\bar{\alpha}}{\partial r_t}\right) = \frac{1}{\sqrt{n}}\left(\frac{\partial\bar{\alpha}}{\partial S_t}\right) \tag{60}$$

where $n$ is the number of equivalent bond-stretching coordinates
comprising $S_t$. The values of the bond-polarizability derivatives
obtained by Yoshino and Bernstein[56] are listed in Table IV, together
with more recent data of Schrötter and Bernstein.[66] Two values are
listed for ethane because two different types of approximations were
introduced in the force field needed for the calculation of $L$ matrix
entries.

The values of $\partial\bar{\alpha}/\partial r_{\text{C-H}}$ listed in Table IV are remarkable for their
constancy. All are about 1.0 Å$^2$ irrespective of the molecular environ-
ment of the CH bonds and the different hybridization states of the C

atoms involved ($sp^3$ for methane, ethane, and neopentane; $sp^2$ for benzene and ethylene; $sp$ for acetylene). This result should be contrasted with the way in which derivatives of the electric moment with respect to C—H bond length, $\partial\mu/\partial r_{C-H}$, vary from one molecule to another, and even from one symmetry species to another of the same molecule.[67,68] The values of $\partial\mu/\partial r_{C-H}$ are derived from intensities of infrared bands. In addition, the data in Table IV show an approximate proportionality of the $\partial\bar{\alpha}/\partial r_{C-C}$ values to the multiplicities of the CC bonds involved. Division of the $\partial\bar{\alpha}/\partial r_{C-C}$ values by the CC bond orders (1 for ethane and neopentane, $1\frac{1}{2}$ for benzene, 2 for ethylene, 3 for acetylene) produces approximately $1.0\ \text{Å}^2$ in each case—the value found for $\partial\bar{\alpha}/\partial r_{C-H}$. Hence, the bond-polarizability derivatives evaluated appear to be proportional to the numbers of bonding electrons in CH and CC bonds, though no obvious distinction between $\sigma$- and $\pi$-type bonds is apparent.

Although great care obviously was taken by Yoshino and Bernstein[56] to remove errors and inconsistencies from their bond-polarizability derivatives, the more recent and more reliable values of Schrötter and Bernstein[66] are seen to be as much as 16% higher for the CH bonds and 22% higher for the CC bonds of benzene. The new values for benzene do not produce $1.0\ \text{Å}^2$ as well when divided by the CH and CC bond orders.

The rather large deviations from constancy apparent in the values of $(\partial\bar{\alpha}/\partial r_{C-C})/f$, where $f$ is the CC bond order, especially when the more recent data are also considered, raise doubts about Yoshino and Bernstein's[56] original conclusion that equal contributions to $\partial\bar{\alpha}/\partial r_{C-C}$ values are made by $\sigma$- and $\pi$-type bonding electrons. Certainly, the results of later work by Chantry and Plane[69] on the Raman intensities of totally symmetric vibrations of simple oxyanions indicate that a $\pi$-electron makes a greater contribution to the derived bond polarizability than does a $\sigma$-electron.

Chantry and Plane's data for the two sets of isoelectronic species— $CO_3^{-2}$ and $NO_3^-$, and $PO_4^{-3}$, $SO_4^{-2}$, and $ClO_4^-$—are given in Table V. Their intensities were all measured relative to the $\nu_1(A_1)$ line of $CCl_4$, through use of an internal $ClO_4^-$ standard, the $\nu_1(A_1)$ line of which was measured relative to $CCl_4$ by mixing these two species in alcoholic solution. Additional corrections were applied to their measured intensities for variations in spectrometer response with wavelength. The derived bond polarizabilities reported in Table V are on a scale by which $\partial\bar{\alpha}/\partial r_{C-H}$ has the value $1.00\ \text{Å}^2$ (cf. Yoshino and Bernstein's[56] experimentally determined value of 1.04 Å given in Table IV). Chantry and Plane used equation (8) to obtain $\partial\bar{\alpha}/\partial Q$ values from their Raman intensities and depolarization ratios, the latter being measured with calibrated Polaroid cylinders[6] surrounding the sample tube. Bond-polarizability derivatives were obtained from these $\partial\bar{\alpha}/\partial Q$ values through the relationships

$$XY_4 : \left(\frac{\partial\bar{\alpha}}{\partial Q_1}\right) = \left(\frac{4}{M_Y}\right)^{\frac{1}{2}} \left(\frac{\partial\bar{\alpha}}{\partial r_{X-Y}}\right) \tag{61}$$

and

$$XY_3 : \left(\frac{\partial\bar{\alpha}}{\partial Q_1}\right) = \left(\frac{3}{M_Y}\right)^{\frac{1}{2}} \left(\frac{\partial\bar{\alpha}}{\partial r_{X-Y}}\right) \tag{62}$$

Three factors might be expected to affect bond-polarizability derivatives: the changing distribution of shared electrons, the effect of unshared or "lone pairs" of electrons, and nonbonded atom or group interactions. Woodward and Long's work[5] (discussed earlier) is cited as evidence of the significant influence of the latter two factors, while the trends in $\partial\bar{\alpha}/\partial r_{C-Cl}$ values determined by Long, Milner, and Thomas[70] (2.00 Å$^2$ for $CCl_4$, 1.86 Å$^2$ for $CHCl_3$, 1.55 Å$^2$ for $CH_2Cl_2$) are attributed to the third factor, nonbonded interactions. By studying isoelectronic series of oxyanions, Chantry and Plane hoped to eliminate the effects of unshared electrons and nonbonded interactions, leaving only the effect of the changing distribution of shared (bonding) electrons to explain the trends in bond-polarizability derivatives.

Arguing on the basis of the decreasing basicity in the series $PO_4^{-3}$, $SO_4^{-2}$, and $ClO_4^-$ and the essential electrical neutrality condition postulated by Pauling,[71] Chantry and Plane have interpreted

**TABLE V**
**Bond-Polarizability Derivatives $\partial\bar{\alpha}/r_{XY}$ (Å$^2$) for Some Oxyanions[69]**

| Species | $\nu_1(A_1)$, cm$^{-1}$ | Relative intensity, $I$ | $(\partial\bar{\alpha}/\partial r_{XY})$, Å$^2$ |
|---------|------------------------|-------------------------|------------------------------------------------|
| $CCl_4$ | 459 | 1.00 (standard) | 2.00* |
| $CO_3^{-2}$ | 1065 | 0.178 ($\rho = 0.106$) | 1.04 |
| $NO_3^-$ | 1040 | 0.500 ($\rho = 0.170$) | 1.64 |
| $PO_4^{-3}$ | 935 | 0.188 | 0.91 |
| $SO_4^{-2}$ | 981 | 0.396 | 1.37 |
| $ClO_4^-$ | 935 | 0.675 | 1.73 |

* From ref. 70.

## TABLE VI
### Bond Polarizability Derivatives for Some X—O Bonds*

| Species | $\partial\bar{\alpha}/\partial r_{X-O}$, $\text{Å}^2$ |
|---------|---------------------|
| $ClO_3^-$ | 1.39 |
| $BrO_3^-$ | 1.95 |
| $IO_3^-$ | 2.24 |
| $IO_4^-$ | 2.74 |
| $H_5IO_6$ | 1.62 |

* X denotes a halogen atom.

their results as indicating the formation of an average of two Cl=O double bonds in the $ClO_4^-$ ion, no double bonds in $PO_4^{-3}$ and an intermediate degree of double bonding in $SO_4^{-2}$. An examination of the symmetry properties of the $p$ and $d$ orbitals involved in $\pi$-bond formation in these tetrahedral oxyanions shows that two is the maximum number of double bonds possible.[72] Further, similar considerations applied to the interpretation of $\partial\bar{\alpha}/\partial r_{X-O}$ values for $CO_3^{-2}$ and $NO_3^-$ lead to the conclusion that an average of one N=O double bond exists in $NO_3^-$, and that $CO_3^{-2}$ probably contains only single $\sigma$-bonds. An equivalent, and perhaps preferable, way of stating these results is that the bond orders or multiplicities in the oxyanions studied are approximately as follows: $ClO_4^-$, $1\frac{1}{2}$; $SO_4^{-2}$, $1\frac{1}{4}$; $PO_4^{-3}$, 1; $NO_3^-$, $1\frac{1}{3}$; $CO_3^{-2}$, 1. Each X atom in these $XY_4$ and $XY_3$ ions is thereby given a formal +1 charge. Chantry and Plane's[69] results imply that the contribution to $\partial\bar{\alpha}/\partial r_{XY}$ from $\pi$-bonds is some 50% higher than for $\sigma$-bonds in the $XY_4$ series, and about 80% higher in the $XY_3$ series. Yoshino and Bernstein's[56] results for hydrocarbons are not necessarily inconsistent with this idea of a larger contribution to $\partial\bar{\alpha}/\partial r$ from $\pi$-electrons than from $\sigma$-electrons, for CC bond shortening with increase in bond multiplicity probably results in a decreased $\sigma$-contribution[73] in the series ethane, ethylene, acetylene, necessitating a larger $\pi$-contribution to account for the direct proportionality between $\partial\bar{\alpha}/\partial r_{CC}$ and CC bond order. It should be noted that a strong dependence of $\partial\bar{\alpha}/\partial r$ values on bond length is implicit in all this.

In a later paper,[74] Chantry and Plane used the following relationship between bond-polarizability derivative $\partial\bar{\alpha}/\partial r_{X-Y}$ and bond order $n$ to account for their results with $ClO_4^-$, $SO_4^{-2}$, and $PO_4^{-3}$:

$$\left(\frac{\partial\bar{\alpha}}{\partial r_{X-Y}}\right) = 0.91 + (n-1)1.64 \tag{63}$$

This relationship was then further employed to predict a value of $\partial\bar{\alpha}/\partial r_{Cl-O}$ for the $ClO_3^-$ ion of $1.46\ \text{Å}^2$, assuming a ClO bond order of $1\frac{1}{3}$, in order to satisfy the essential electrical neutrality principle. The good agreement of this with their experimentally determined value of $1.39\ \text{Å}^2$ led Chantry and Plane into an interpretation of their further results for the ions $BrO_3^-$, $IO_3^-$, and $IO_4^-$ and for $H_5IO_6$. These results are summarized in Table VI.

Within experimental error, the oxyiodide bond-polarizability derivatives obey the relation

$$\left(\frac{\partial\bar{\alpha}}{\partial r_{I-O}}\right) = 1.60 + (n-1)\ 2.10 \tag{64}$$

where the bond order $n$ is determined by assuming a $+1$ charge on the I atoms in each case. The increased contributions of both the $\sigma$-bonds (1.60) and $\pi$-bonds (2.10) in oxyiodides, as compared with the P, S, and Cl oxyanions [see Table V and equation (63)], are in line with Woodward and Long's[5] findings for the Group IV tetrahalides (see Table I and earlier discussion). The $H_5IO_6$ was treated as $IO_6^{-5}$ for this comparison. In conclusion, Chantry and Plane's work on oxyanions appears to provide good evidence for a degree of bond multiplicity consistent with the retention of a formal $+1$ charge on the central atom in all the species studied.

Continuing their investigations of chemical bonding through the determination of Raman intensities, Chantry and Plane next

### TABLE VII
### CN Bond-Polarizability Derivatives for Some Complex Metal Cyanides[75]

| Cyanide | $A_1$ frequencies, cm$^{-1}$ | $\partial\bar{\alpha}/\partial r_{C-N}$, Å$^2$ |
|---------|------------------------------|------------------------------------------------|
| $CN^-$ | 2079 | 1.71 |
| $Zn(CN)_4^{-2}$ | 2143 | 2.12 |
| $Cd(CN)_4^{-2}$ | 2145 | 2.16 |
| $Hg(CN)_4^{-2}$ | 2148 | 2.16 |
|  | 340 |  |
| $Co(CN)_6^{-3}$ | 2152 | 2.21 |
|  | 404 |  |
| $Cu(CN)_4^{-3}$ | 2094 | 2.84 |
| $Ag(CN)_4^{-3}$ | 2097 | 3.03 |
| $Fe(CN)_6^{-4}$ | 2094 | 3.18 |

turned their attention to the complex cyanides of Zn(II), Cd(II), Hg(II), Co(III), Cu(I), Ag(I), and Fe(II) in an attempt to determine the metal–ligand bond orders. They found[75] the low-frequency M—C (where M is the metal atom) totally symmetric stretching modes too weak to be observed in all cases except those of Hg(II) and Co(III), so that most of the calculations and arguments presented here are based on their symmetric CN stretching band intensities. Table VII lists their derived bond polarizabilities for the C—N bonds in the complexes, all the values being on the scale used in Tables V and VI. Free $CN^-$, in an aqueous NaCN solution, was used as one standard for comparison with the complexes, and $CH_3CN$ as another. Based on the intensities of the four $A_1$ frequencies at 2941, 2248, 1371, and 919 $cm^{-1}$, the bond-polarizability derivatives for the $CH_3CN$ molecule were found to be $\partial\bar{\alpha}/\partial r_{C-H} = 0.86$ Å$^2$, $\partial\bar{\alpha}/\partial r_{C-N} = 2.61$ Å$^2$, and $\partial\bar{\alpha}/\partial r_{C-C} = 0.272$ Å$^2$. The exceptionally low value of the derivative for the CC bond (cf. values in Table IV) lends weight to the argument that as the amount of $s$ character in the carbon hybrid bonding orbitals increases, the value of $\partial\bar{\alpha}/\partial r_{C-C}$ decreases: $CH_3CN$ is presumed to involve $sp^3$—$sp$ bonding. This proposition is in keeping with Chantry and Plane's results for oxyanions, with the extremely low intensity of the central CC stretching mode of diacetylene,[76] and with the failure to detect M—C bands from most of the complex cyanides studied.

The complexes listed in Table VII are divided into two distinct groups by both the frequencies (and hence force constants) and the polarizability derivatives of the CN bonds. $Zn(CN)_4^{-2}$, $Cd(CN)_4^{-2}$, $Hg(CN)_4^{-2}$, and $Co(CN)_6^{-3}$ all have derivatives lying between the values for $CN^-$ and $CH_3CN$, suggestive of partial MC $\sigma$-bonds (without $\pi$-bonding) occurring to an extent consistent with essential electrical neutrality. However, the values of the CN bond-polarizability derivatives for $Cu(CN)_4^{-3}$, $Ag(CN)_4^{-3}$, and $Fe(CN)_6^{-4}$ are all larger than that for $CH_3CN$, which at first sight suggests an MC bond order greater than unity. This interpretation is evidently unreasonable, however, for it implies the removal of bonding electrons from the CN region, which surely should decrease the value of $\partial\bar{\alpha}/\partial r_{CN}$ rather than increase it. Chantry and Plane have interpreted these results as exemplifying a breakdown of the simple bond-polarizability theory, i.e., for these anomalous Cu, Ag, and Fe species, it seems that polarizability changes cannot be localized in bonds. Support for the types of structure proposed by Pauling,[77] wherein involvement of occupied metal $d$-orbitals in back-bonding to the CN ligands results in partial double-bond character of the metal–carbon bonds, is not apparent from the Zn, Cd, Hg, and Co results.

The conclusions from the cyanide intensity work are to be contrasted with those made by Woodward and Creighton[78] and Woodward and Ware[79] on the basis of their work on a group of hexahalide anions of transition elements. Measurements of the relative intensities of the $v_1(A_{1g}$ stretch), $v_2(E_g$ stretch), and $v_5(F_{2g}$ bend) Raman bands from the octahedral $MX_6$ species $PdCl_6^{-2}$, $PtCl_6^{-2}$, $PtBr_6^{-2}$, and $PtF_6^{-2}$ have been interpreted by these authors in terms of $M—X$ $(d—d)$ $\pi$-bonding, involving filled metal $d$-orbitals as donors and vacant ligand $d$-orbitals as acceptors. The contrast with the conclusions from the cyanide work is striking, since cyanide complexes are generally considered to be more extensively $\pi$-bonded than halide complexes.

Table VIII lists the intensity data[78,79] for the hexahalide species, together with the frequencies of the exciting lines used $(v_0)$ and the frequencies of the lowest-lying excited electronic states $(v_e)$ to which *allowed* transitions can be made.[80] The latter are to be used in a later part of this discussion. Figures for the non–transition metal hexahalide $SnCl_6^{-2}$ are included for comparison.[81] The spectra were recorded photographically, and intensities were estimated by eye. The various exciting frequencies were obtained from an AC helium lamp (yellow line at 17,000 cm$^{-1}$, red line at 15,000 cm$^{-1}$) and the usual type of mercury arc (22,900 cm$^{-1}$, green line at 18,300 cm$^{-1}$). The assignments of Raman bands to $A_{1g}$, $E_g$, and $F_{2g}$ species were made unambiguously on the basis of frequencies and degrees of depolarization. The same pattern of relative band intensities in the $PtCl_6^{-2}$ spectrum excited by three different frequencies (17,000 cm$^{-1}$, 15,000 cm$^{-1}$, and 22,900 cm$^{-1}$) was used as evidence for the absence of a resonance Raman effect.

## TABLE VIII
### Raman Data for Hexahalide Complexes

| Parameters | $PdCl_6^{-2}$ | $PtCl_6^{-2}$ | $PtBr_6^{-2}$ | $PtF_6^{-2}$ | $SnCl_6^{-2}$ |
|---|---|---|---|---|---|
| $I_1(A_{1g})$ | 10 | 10 | 10 | 10 | 10 |
| $I_2(E_g)$ | 14 | 14 | 14 | 3 | 2 |
| $I_5(T_{2g})$ | 10 | 9 | 9 | 2 | 4 |
| $v_0$(cm$^{-1}$) | 15,000 | 17,000 | 15,000 | 18,300 | 22,900 |
| $v_e$(cm$^{-1}$) | 29,400 | 38,200 | 31,800 | >60,000 | 44,900 |
| $v_e - v_0$(cm$^{-1}$) | 14,400 | 21,200 | 16,800 | >41,700 | 22,000 |

Raman spectra of a large number of octahedral $MX_6$-type compounds have been investigated in the past (see Jorgensen[80] for examples and further references), and all have shown the relative intensity pattern $I_1 \gg I_2, I_5$. The spectra from $PdCl_6^{-2}$, $PtCl_6^{-2}$, and $PtBr_6^{-2}$ are remarkable for their large deviations from this "normal" pattern. An obvious difference between the hexahalides and the "normal" $SnCl_6^{-2}$ is the presence of six $t_{2g}$ $d$-electrons on the central metal atom. Woodward and Ware[79] have proposed that these electrons are delocalized by overlapping with vacant $d$-orbitals on Cl and Br ligands. The resultant type of $(d-d)$ $\pi$-bond formation is not possible with F as a ligand, because of the unavailability of low-energy $d$-orbitals; hence the "normal" pattern of intensities from $PtF_6^{-2}$.

In terms of the derivatives of the bond-polarizability components $\alpha'_e$ and $\alpha'_p$ (where $\alpha'_e = \partial\bar{\alpha}_e/\partial r$, the derivative of the component along the bond, and $\alpha'_p = \partial\bar{\alpha}_p/\partial r$, the derivative of the component perpendicular to the bond), the ratio of the $v_1$ and $v_2$ band intensities may be expressed as[78]

$$\frac{I_1}{I_2} = \frac{5A_1[1 + (2\alpha'_p/\alpha'_e)]^2}{13A_2[1 - (\alpha'_p/\alpha'_e)]^2} \tag{65}$$

where $A_i = (v_0 - \Delta v_i)^4/\Delta v_i[1 - \exp(-h\Delta v_i/kT)]$. The expression is based on the Wolkenstein theory,[13] discussed in the second section of this chapter. Calculations based on this have yielded[78,79] values of $\alpha'_p/\alpha'_e = 0.55$ for $SnCl_6^{-2}$, 0.39 for $PtF_6^{-2}$, and 0.12 for $PtCl_6^{-2}$ suggesting an inverse proportionality to bond order. This conflicts directly with Chantry and Plane's conclusion, based on $CH_4$, $CCl_4$, and $CN^-$ Raman intensities, that $\alpha'_p/\alpha'_e$ increases with bond order.[82] However, Woodward and Ware[79] have warned that the Wolkenstein assumptions underlying this treatment may not be valid in all cases, as demonstrated by Chantry and Plane's cyanide results discussed earlier.

An alternative approach to the explanation of the anomalous intensity patterns in the spectra of $PdCl_6^{-2}$, $PtCl_6^{-2}$, and $PtBr_6^{-2}$ has been taken by Albrecht and Taylor.[83] Although the values for $(v_e - v_0)$ listed in Table VIII indicate that in no case can the condition be properly described as a near-resonance one, a particularly selective vibronic activity of the $E_g$ modes in the low-lying electronic states of the hexahalide ions could account for the high relative intensities of these non-totally symmetric modes found in the Raman spectra. Application of the vibronic theory of Raman intensities, as developed by Albrecht[17] and discussed briefly in the first section of this chapter, has shown that as the resonance condition is approached all three Raman active modes should increase in intensity for $d^{10}$-complexes of the $MX_6$-type, but not for $d^6$-complexes. In this latter case, only the non–totally symmetric modes should be affected. The agreement of these predictions with the data of Table VIII, wherein $SnCl_6^{-2}$ is a $d^{10}$-example, the others being $d^6$-species, appears still more convincing when taken together with the fact that the conditions used for exciting the "normal" $PtF_6^{-2}$ spectrum were much farther from resonance than in any other case. Under these conditions, the selective activity of the non-totally symmetric modes in the low-energy states would be much less important in the context of the summation over all states used in the equations for Raman intensity.[17]

In concluding this section on bond orders, it should be noticed that only relatively simple molecular species have been discussed. The uncertainty in Raman band assignments, mixing of different types of vibrational motion in a given normal mode, and the practical problem of resolving overlapping bands are factors which combine to make meaningful interpretations of intensities from complex molecules difficult. Hester and Plane's[84] work on the trisoxalato and trisacetylacetonato complexes of aluminum, gallium, and indium demonstrates the difficulties, while showing the conditions under which useful information can be obtained from Raman intensities of large molecules (e.g., aluminum trisacetylacetonate has 43 atoms per molecule). These oxalato and acetylacetonato species of $D_3$ symmetry each give a strong, sharp, and highly polarized line of low frequency, which appears to be characteristic of the totally symmetric $MO_6$ stretching mode. Comparison of these line intensities within and between the two series of complexes has provided support for the idea of electron delocalization or pseudo-aromatic resonance in the Group III metal–acetylacetonate chelate rings. However, most of the work done with Raman intensities from large molecules has been concerned with establishing values for the molar intensities of bands characterizing functional groups in organic molecules. Behringer and Brandmüller[23] have reviewed this work in some detail, showing that, provided the excitation frequency is far removed from resonance, characteristic molar intensities can be assigned to such groups as $C=C$, $C—Cl$, $C\equiv N$, and even $CH_2$. Moreover, with the same provision as above, simple additivity rules are found to be applicable to $C—H$ and

C=C groups in large organic molecules, further demonstrating the utility of Raman intensities. For mono- and disubstituted benzenes, Venkateswarlu and Radhakrishnan[85] have been able to correlate variations in characteristic group intensities with the nature of the substituents.

## INTERMOLECULAR INTERACTIONS IN LIQUIDS

In the previous sections of this chapter, Raman intensity data obtained from gases, liquids, and dissolved species in liquid solutions all have been used to provide information on the nature of chemical bonding. An absolute scale of intensities and hence of derived bond polarizabilities determined for gas-phase species[56] has even been used for aqueous electrolyte solution data.[69] Yet it is clear that the formulas of Placzek's theory[2] for the intensities of Rayleigh and Raman scattering are strictly applicable only to gases. How does this affect what has gone before?

Brandmüller and Schrötter[86] have considered the effect on the Rayleigh intensity of intermolecular interactions and of interference of the scattered light caused by short-range ordering of molecules in the liquid phase. Their comparison of Placzek's formulas with those derived from the fluctuation theory of Smoluchowski[87] and Einstein,[86] taken with Carpenter and Krigbaum's[89] review of light-scattering data, has led to an estimated possible error of $\pm 30\%$ in liquid-phase intensities. Even this large an error is still insufficient to account for the discrepancy between the intensities of the $CCl_4$ 459 cm$^{-1}$ band measured relative to the Rayleigh line in the gas and liquid phases. The 459 cm$^{-1}$ band intensity from gaseous $CCl_4$ gives a value[90] of $45(\bar{\alpha}')^2 = 22 \times 10^{-32} \times N$ cm$^4$/g, where $N$ is Avogadro's number. Similar measurements with liquid $CCl_4$[86,91] give $(56 \pm 15) \times 10^{-32} \times N$ cm$^4$/g. Thus, the gas value, while agreeing quite well with Whiffen's[92] calculated value of $28 \times 10^{-32} \times N$ cm$^4$/g, is smaller than the liquid value by a factor of approximately two. This 459 cm$^{-1}$ $v_1(A_1)$ band of $CCl_4$ is extensively used as an internal and external standard, so that it becomes important to understand the reasons for this discrepancy. Since the doubling of the $CCl_4$ band intensity in the liquid phase evidently results from some form of intermolecular interaction, it is of interest to examine the effect in more detail, and to make similar gas/liquid comparisons for other substances.

## TABLE IX
## Comparison of Standard Intensities from Gas and Liquid Phases of Substances[66]

| Substances | $\Delta v$, cm$^{-1}$ | $S$ (gas) | $S$ (liquid) | Source references |
|---|---|---|---|---|
| Carbon tetrachloride | 459 | 1.00 | 1.00 | |
| | 314 | 0.36 | 0.32 | 10, 70 |
| Chloroform | 671 | 0.59 | 0.60 | 10, 70 |
| | 365 | 0.3 | 0.28 | 10, 70 |
| Benzene | 3070 | 12.7 | | |
| | 3056 | 15 | 21 | 10, 95 |
| | 991 | 4.45 | 4.1 | 10, 94, 95 |
| Acetonitrile | 2942 | 6.8 | 4.7 | 96 |
| | 2249 | 3.2 | 1.7 | 96 |
| | 918 | 0.3 | 0.13 | 96 |
| Carbon disulfide | 658 | 3.7 | 5.39 | 94 |
| | 802 | | | |
| Methanol | $\Sigma v_{CH}$ | 9.5 | 6.9 | 94 |
| n-Pentane | $\Sigma v_{CH}$ | 42 | 36.2 | 94 |
| n-Hexane | $\Sigma v_{CH}$ | 50 | 40.6 | 94 |

Schrötter and Bernstein[66] have assembled data on standard intensities from a variety of substances in the liquid phase for comparison with their measured standard intensities from the same substances in the gas phase. Their data are reproduced in Table IX. Standard (integrated) intensities $S$ are all on the scale proposed by Bernstein and Allen,[10] which uses the 459 cm$^{-1}$ line of CCl$_4$ as a standard, namely,

$$S = \frac{I_{obs}}{I_{459}} \frac{1 + \rho_{459}}{1 + \rho_{obs}} \frac{n^2}{n^2_{CCl_4}} \frac{\sigma_{\Delta v}}{\sigma_{459}} \frac{R(n)}{R_{CCl_4}} \frac{M}{d} \left(\frac{d}{M}\right)_{CCl_4}$$

$$\times \frac{\Delta v}{459} \left(\frac{v_0 - 459}{v_0 - \Delta v}\right)^4 \left[\frac{1 - \exp(-1.44\Delta v/T)}{1 - \exp(-1.44 \times 459/T)}\right]$$

$$S = \frac{g_{\Delta v}[45(\bar{\alpha}')^2 + 7(\gamma')^2]_{\Delta v}}{[45(\bar{\alpha}')^2 + 7(\gamma')^2]_{459}} \tag{66}$$

Here $n$ represents refractive index, $g_{\Delta v}$ is the degree of degeneracy of the mode, $\sigma$ is the spectral sensitivity of the phototube, $R$ is the reflection loss, $M$ is the molecular weight, and $d$ is the density. Equation (66) is based on equation (12) given earlier. Thus, these standard intensities $S$ are fully corrected for those factors which may vary from one measurement to another. The refractive-index correction $n^2 R(n)$ represents the least well-established correction factor.[66] Rea has examined this factor in great detail.[93] The value of $S$ for the 459 cm$^{-1}$ line of $CCl_4$ is taken as 1.00 for both liquid and gas, although, as stated above, it should be remembered that there is a factor of 2 intensity enhancement for this line in the liquid.

It is seen that there are wide variations in behavior from one liquid to another, although in most cases the ratio of liquid intensity to gas intensity is greater than 0.5, i.e., intensities are usually increased in the liquid. Acetonitrile is an exception to this. The influence of intermolecular inter ctions on the intensities of bands evidently varies considerably from one substance to another, and even from one vibrational mode to another in the same substance. Of the data in Table IX, that for acetonitrile ($CH_3CN$) shows this latter variation most strikingly; the bands at 918, 2249, and 2942 cm$^{-1}$ are evidently progressively more affected by the intermolecular field in the liquid.

Further examination of the corrections to be applied to measured Raman intensities from liquids has led to a preliminary formulation[97] of the mechanism of intermolecular interactions in terms of van der Waals forces and $\pi$-electron field effects. Intensity measurements with $CCl_4$–hexane liquid mixtures give data in agreement with values calculated on the basis of a potential–energy function using dispersion forces only to describe the intermolecular field, but for $CS_2$–benzene mixtures such a field is evidently inadequate, and stronger interactions involving the aromatic $\pi$-electrons must be considered.[97] Bernstein[97] has shown how the extensive data of Pivovarov[98] and Rea[99] can be used to provide this type of information from solvent effects on Raman intensities, although Rea[99] had previously concluded that current theories were inadequate to provide a satisfactory explanation for his observations. These and other recent investigations of solvent effects[100–103] indicate that further examination of Raman intensity and depolarization ratios may enable great progress to be made in the near future in understanding the nature of the interactions between molecules in liquids. However, even in the area of interpretation of intensities from simple gaseous substances, much remains to be done in terms of the nature of their chemical bonding, as is well illustrated by recent results[104] from gaseous $CH_4$, $CD_4$, $CF_4$, $SiF_4$, $SF_6$, $SeF_6$, and $TeF_6$. Evidently, great caution must be exercised in the evaluation of intensities, particularly from liquids, in terms of chemical bonding. While intensity differences within a series of similar compounds studied under similar conditions often can be interpreted in terms of trends in the nature of the chemical bonding involved, the exact interpretation of single intensity values from liquids is less certain. Due to the square-root relationship, $\bar{\alpha}' \propto \sqrt{I}$, the approximate doubling of intensities by condensation of substances from gas to liquid state is less serious in terms of change in $\bar{\alpha}'$ values, but remains an important complication. The formidable problem of conversion of measured intensities to an absolute scale is reasonably well understood now, though, and the future for these kinds of investigations looks promising.

## REFERENCES

1. L. Pauling and E. B. Wilson, *Introduction to Quantum Mechanics*, McGraw-Hill Book Company, New York (1935).
2. G. Placzek, *Handbuch der Radiologie, Volume VI*, Part 2, Leipzig, 1934, p. 205.
3. E. B. Wilson, J. C. Decius, and P. C. Cross, *Molecular Vibrations*, McGraw-Hill Book Company, New York, 1955.
4. R. W. Wood, *Phys. Rev.* **36**: 1421 (1930).
5. L. A. Woodward and D. A. Long, *Trans. Faraday Soc.* **45**: 1131 (1949).
6. J. T. Edsall and E. B. Wilson, *J. Chem. Phys.* **6**: 124 (1938).
7. D. H. Rank and R. E. Kagarisse, *J. Opt. Soc. Am.* **40**: 89 (1950).
8. A. E. Douglas and D. H. Rank, *J. Opt. Soc. Am.* **38**: 281 (1948).
9. J. Brandmüller and H. Moser, *Einführung in die Raman Spektroskopie*, Steinkopff, Darmstadt, 1962.
10. H. J. Bernstein and G. Allen, *J. Opt. Soc. Am.* **45**: 237 (1955).
11. W. H. Martin, *Trans. Roy. Soc. Can. Sect. III* **17**: 151 (1923).
12. M. Wolkenstein, *Compt. Rend. Acad. Sci. URSS* **30**: 791 (1941).
13. M. Eliashevich and M. Wolkenstein, *J. Phys. USSR* **9**: 101, 326 (1945).
14. D. A. Long, *Proc. Roy. Soc. (London)* **A 217**: 203 (1953).
15. D. A. Long, A. H. S. Matterson, and L. A. Woodward, *Proc. Roy. Soc. (London)* **A 224**: 33 (1954).
16. K. A. Taylor and L. A. Woodward, *Proc. Roy. Soc. (London)* **A 264**: 558 (1961).
17. A. C. Albrecht, *J. Chem. Phys.* **34**: 1476 (1961).
18. J. H. Van Vleck, *Proc. Nat. Acad. Sci. U.S.* **15**: 754 (1929).
19. G. Hertzberg and E. Teller, *Z. Physik. Chem. (Leipzig)* **B 21**: 410 (1933).
20. A. C. Albrecht, *J. Chem. Phys.* **33**: 156 (1960).
21. P. P. Shorygin, *J. Phys. Chem (USSR)* **21**: 1125 (1947).
22. P. P. Shorygin, *Izv. Akad. Nauk SSSR, Ser. Fiz.* **12**: 576 (1948).
23. J. Behringer and J. Brandmüller, *Z. Elektrochem.* **60**: 643 (1956).
24. D. G. Rea, *J. Mol. Spectry.* **4**: 499 (1960).
25. E. B. Wilson, *J. Chem. Phys.* **7**: 1047 (1939); **9**: 76 (1941).
26. E. B. Wilson, J. C. Decius, and P. C. Cross, *Molecular Vibrations*, McGraw-Hill Book Company, New York, 1955.
27. B. L. Crawford and W. H. Fletcher, *J. Chem. Phys.* **19**: 141 (1951).
28. D. N. Walters and L. A. Woodward, *Proc. Roy. Soc. (London)* **A 246**: 119 (1958).
29. G. W. Chantry and L. A. Woodward, *Trans. Faraday Soc.* **56**: 1110 (1959).
30. D. A. Long, R. B. Gravenor, and D. C. Milner, *Trans. Faraday Soc.* **59**: 46 (1963).
31. H. L. Welsh, M. F. Crawford, and G. D. Scott, *J. Chem. Phys.* **16**: 97 (1948).

32. I. Hansen-Damaschun, *Z. Physik. Chem.* **B 22**: 97 (1933).
33. L. Pauling, *Nature of the Chemical Bond*, 3rd. ed., Cornell University Press, Ithaca, N.Y., 1960, p. 98.
34. G. Placzek, *Handbuch der Radiologie, Volume VI*, Part 2, Leipzig, 1934, p. 366.
35. G. Placzek, *Z. Physik* **70**: 84 (1931).
36. J. P. Mathieu and M. Lounsbury, *Discussions Faraday Soc.* **9**: 196 (1950).
37. J. R. Ferraro, *J. Mol. Spectry.* **4**: 99 (1960).
38. R. E. Hester and R. A. Plane, *Inorg. Chem.* **3**: 769 (1964).
39. R. E. Hester and R. A. Plane, *J. Chem. Phys.* **40**: 411 (1964).
40. S. C. Wait and G. J. Janz, *Quart. Rev. (London)* **17**: 225 (1963).
41. J. K. Wilmshurst and S. Senderoff, *J. Chem. Phys.* **35**: 1078 (1961).
42. J. H. B. George, J. A. Rolfe, and L. A. Woodward, *Trans. Faraday Soc.* **49**: 375 (1953).
43. R. P. Bell and J. H. B. George, *Trans. Faraday Soc.* **49**: 619 (1953).
44. P. L. Goggin and L. A. Woodward, *Trans. Faraday Soc.* **56**: 1591 (1960).
45. H. Lee and J. K. Wilmshurst, *Australian J. Chem.* **17**: 943 (1964).
46. R. A. Robinson and R. H. Stokes, *Electrolyte Solutions*, 2nd ed., Butterworths, London, 1959, Chap. 14.
47. C. W. Davies, in: W. J. Hamer (ed.), *The Structure of Electrolytic Solutions*, John Wiley & Sons, Inc., New York, 1959, Chap. 3.
48. P. L. Goggin and L. A. Woodward, *Trans. Faraday Soc.* **58**: 1495 (1962).
49. R. E. Hester, R. A. Plane, and G. E. Walrafen, *J. Chem. Phys.* **38**: 249 (1963).
50. R. E. Hester and R. A. Plane, *Inorg. Chem.* **3**: 768 (1964).
51. G. Hertzberg, *Molecular Spectra and Molecular Structure II*, D. Van Nostrand Company, Inc., Princeton, N.J., 1945.
52. R. E. Hester, Thesis, Cornell University, 1962.
53. P. C. Cross, J. Burnham, and P. A. Leighton, *J. Am. Chem. Soc.* **59**: 1134 (1937).
54. A. da Silveira, M. A. Marques, and N. M. Marques, *Compt. Rend.* **252**: 3983 (1961).
55. J. P. Mathieu, *Compt. Rend.* **231**: 896 (1950); see also R. L. Lafont, *Compt. Rend.* **244**: 1481 (1957).
56. T. Yoshino and H. J. Bernstein, *Spectrochim. Acta* **14**: 127 (1959).
57. T. Yoshino and H. J. Bernstein, *J. Mol. Spectry.* **2**: 213 (1958).
58. H. L. Welsh, M. F. Crawford, T. R. Thomas, and G. R. Love, *Can. J. Phys.* **30**: 577 (1952).
59. G. Hertzberg, *Spectra of Diatomic Molecules*, D. Van Nostrand Company, Inc., New York, 1950. p. 128.
60. B. L. Crawford, Jr., and F. A. Miller, *J. Chem. Phys.* **17**: 249 (1949).
61. B. L. Crawford, Jr., J. A. Lancaster, and R. G. Inskeep, *J. Chem. Phys.* **21**: 678 (1963).
62. G. E. Hansen and D. M. Dennison, *J. Chem. Phys.* **20**: 313 (1952).
63. T. Simanouchi, *J. Chem. Phys.* **17**: 849 (1949).
64. D. A. Long, *Proc. Roy. Soc. (London)* **A 224**: 33 (1954).
65. J. W. Linnett, *J. Chem. Phys.* **8**: 91 (1940).
66. H. W. Schrötter and H. J. Bernstein, *J. Mol. Spectry.* **12**: 1 (1964).
67. R. C. Golike, I. M. Mills, W. B. Person, and B. L. Crawford, Jr., *J. Chem. Phys.* **25**: 1266 (1956).
68. I. M. Nyquist, I. M. Mills, W. B. Person, and B. L. Crawford, Jr., *J. Chem. Phys.* **26**: 552 (1957).
69. G. W. Chantry and R. A. Plane, *J. Chem. Phys.* **32**: 319 (1960).
70. D. A. Long, D. C. Milner, and A. G. Thomas, *Proc. Roy. Soc. (London)* **A 237**: 197 (1956).
71. L. Pauling, *J. Chem. Soc.* 1461 (1948).
72. H. Eyring, J. Walter, and G. E. Kimball, *Quantum Chemistry*, John Wiley & Sons, Inc., New York, 1944, p. 231.

73. R. P. Bell, *Trans. Faraday Soc.* **38**: 422 (1942).
74. G. W. Chantry and R. A. Plane, *J. Chem. Phys.* **34**: 1268 (1961).
75. G. W. Chantry and R. A. Plane, *J. Chem. Phys.* **35**: 1027 (1961).
76. A. V. Jones, *Proc. Roy. Soc. (London)* **A 211**: 285 (1952).
77. L. Pauling, *Nature of the Chemical Bond*, 3rd ed., Cornell University Press, Ithaca, N.Y., 1960, p. 337.
78. L. A. Woodward and J. A. Creighton, *Spectrochim. Acta* **17**: 594 (1961).
79. L. A. Woodward and M. J. Ware, *Spectrochim. Acta* **19**: 775 (1963).
80. C. K. Jorgensen, *Mol. Phys.* **2**: 309 (1959).
81. L. A. Woodward and L. E. Anderson, *J. Chem. Soc.* 1284 (1957).
82. G. W. Chantry and R. A. Plane, *J. Chem. Phys.* **33**: 634 (1960).
83. A. C. Albrecht and K. A. Taylor, paper presented at the VIIth European Congress on Molecular Spectroscopy, Budapest, Hungary, July 1963 (to be published).
84. R. E. Hester and R. A. Plane, *Inorg. Chem.* **3**: 513 (1964).
85. K. Venkateswarlu and M. Radhakrishnan, *Spectrochim. Acta* **18**: 1433 (1962).
86. J. Brandmüller and H. Schrötter, *Z. Physik* **149**: 131 (1957).
87. M. Smoluchowski, *Ann. Physik* **25**: 205 (1908).
88. A. Einstein, *Ann. Physik* **33**: 1275 (1910).
89. D. K. Carpenter and W. R. Krigbaum, *J. Chem. Phys.* **24**: 1041 (1956).
90. H. W. Schrötter and H. J. Bernstein, *J. Mol. Spectry.* **7**: 464 (1961).
91. H. W. Schrötter, *Z. Elektrochem.* **64**: 853 (1960).
92. D. G. Whiffen, *J. Opt. Soc. Am.* **47**: 568 (1957).
93. D. G. Rea, *J. Opt. Soc. Am.* **49**: 90 (1959).
94. H. J. Bernstein and J. A. Koningstein (to be published).
95. G. Allen and H. J. Bernstein, *Can. J. Chem.* **33**: 1137 (1955).
96. J. C. Evans and H. J. Bernstein, *Can. J. Chem.* **33**: 1746 (1955).
97. H. J. Bernstein, *Pure Appl. Chem.* **4**: 23 (1962).
98. V. M. Pivovarov, *Opt. Spectr. (USSR) (English Transl.)* **9**: 139 (1960).
99. D. G. Rea, *J. Mol. Spectry.* **4**: 507 (1960).
100. P. R. Ryason, *J. Mol. Spectry.* **8**: 579 (1962).
101. I. J. P. Jesson and H. W. Thompson, *Proc. Roy. Soc.* **A 268**: 68 (1962).
102. S. A. Tare and H. W. Thompson, *Spectrochim. Acta* **18**: 1095 (1962).
103. G. W. Chantry, *Spectrochim. Acta* **21**: 1007 (1965).
104. D. A. Long and E. L. Thomas, *Trans. Faraday Soc.* **59**: 1026 (1963).

*Chapter 5*

# Ionic Melts

## G. J. Janz and S. C. Wait, Jr.

*Department of Chemistry*
*Rensselaer Polytechnic Institute*
*Troy, New York*

## INTRODUCTION

A comprehensive survey of the various aspects of fused-salts studies and the current status of knowledge is given in the two research-level monographs[1,2] recently published. It is clear that the vibrational spectroscopy of the inorganic molten state is still relatively little explored and that the factors contributing to structural aspects in this class of liquids are far from clearly defined.

From a correlation of the widely scattered data for fused salts in the fields of electrical conductivity, transport, and viscosity, it is clear that most salts fall conveniently into three broad groups: (1) single-salt melts with $E_\eta > E_\Lambda$;* (2) single-salt melts with $E_\eta = E_\Lambda$; and (3) single-salt melts with $E_\eta < E_\Lambda$. The background of this classification has been given elsewhere[3] in detail, together with the results for some 126 inorganic compounds as single-salt melts. It is sufficient for the present to note that the relatively high-melting inorganic salts $(E_\eta > E_\Lambda)$ appear best suited for studies of environmental factors and interparticulate interactions ("partial covalency"). The use of salts having polyatomic cationic or anionic components provides ready spectroscopic "detectors," since it should be possible to correlate the transgression of spectroscopic selection rules with the type and intensity of nearest-neighbor interactions in the molten salts. Nitrates, carbonates, sulfates, thiocyanates, and perchlorates belong to this group. For investigations of steric factors and packing effects in fused salts, the second group $(E_\eta = E_\Lambda)$ appears most promising. Such salts have

---

* $E_\eta$ and $E_\Lambda$ are the energies of activation for viscous flow and electrical conductance, respectively.

relatively low melting points (e.g., $< 100°C$ for $AlBr_3$ and quaternary n-amyl ammonium thiocyanate). The third class ($E_\eta < E_\Lambda$) embraces salts that form predominantly molecular melts (e.g., mercuric halides and inorganic polymers). These appear well suited for spectroscopic studies of ion–dipole, ion–quadrupole, and ligand–field interactions in the inorganic molten state.

The vibrational spectroscopy of aqueous ionic solutions at ambient temperatures also relates to fused-salt spectroscopy, since solutes in the ultraconcentrated concentration range frequently mimic the structure of the salts as melts. Comparison with characteristic frequencies in aqueous systems, in which the nature of the polyatomic solute species is unambiguously known, has been a semiempirical principle in the assignment of melt spectra.

In this chapter, some of the problems encountered in the practical application of Raman spectroscopy to the study of ionic melts are first considered. Features of experimental work indicating differences from conventional practice are stressed. The vibrational spectroscopy of inorganic salts in the molten state is the main part of this chapter. The results of infrared studies on ionic melts are also given.

## TECHNIQUES

The apparatus and containers in contact with inorganic melts should be suitable for use at elevated temperatures and inert to attack by salt melts.[4] Pyrex glass or quartz have been widely used for investigations at temperatures up to 550 and 1400°C, respectively. The known and small thermal expansivities of these materials are frequently additional advantages. Attack of these materials is often dependent on the purity of the melt. Most nitrates and halides, lithium salts excepted, do not attack glass or quartz if the salts are pure; the presence of water or hydroxide impurities may lead to severe corrosion. At high temperatures, Pyrex glass in contact with fused salts and other glasses exhibits ion-exchange properties, a factor of possible concern for studies with dilute solutions, or if sodium ions interfere. Molten fluorides, hydroxides, and carbonates are among those which are chemically too reactive for glass or quartz container materials. Nickel has been used successfully with molten fluorides. The corrosion chemistry in chloride and carbonate melts has recently been considered from the thermodynamic viewpoint by Littlewood[5] and Ingram and Janz.[6]

Where Pyrex or quartz are unsuitable, the noble metals or their alloys have been widely used for apparatus design. Refractory metals, such as molybdenum, tungsten, and tantalum, are also used; fabrication techniques for the latter metals are still essentially laboratory arts.[7,8] Pressed boron nitride, synthetic sapphire, alumina, and magnesium oxide have been used for the more "reactive" molten salts. Information on the properties of such materials and fabrication techniques have been reviewed by Livey[9] and Campbell.[10] The summaries[3,4] of empirical observations for melts in contact with a wide range of materials also provide useful guides to the selection of container materials.

Purification of melts is an important consideration. Troublesome impurities are water and heavy metals. For the alkali or alkaline earth salts, lithium and magnesium salts excepted (i.e., not strongly hydrated salts), water is readily removed by controlled heating *in vacuo*. Thermo-gravimetric weight-loss studies for LiCl, NaCl, and KCl have been reported[11] from room temperature to temperatures well above melting; the latter two salts can be vacuum-dried at moderately high temperatures without elaborate precautions. More strongly hydrated salts, especially LiCl, $MgCl_2$, and probably $CaCl_2$, require additional care since hydrolysis may occur. The hydroxides thus formed can react with other solute species in the melts and also can lead to glass attack. For the purification of the melts containing LiCl, the best procedure is that of Laitinen[12] and of Hill, Perano, and Osteryoung.[13] The mixture, pre-dried at moderate temperatures, is continuously purged with anhydrous HCl while the temperature is gradually raised until fusion is achieved. An argon flush is then used to sweep out the HCl. Heavy-metal impurities may be removed at this stage by the addition of magnesium or by electrolysis. Filtration of the molten mixture is recommended as a final step in the purification procedure. An assembly for the purification of 100–200 g quantities of salt has been described by Hill;[13] a similar arrangement for smaller quantities is described by Boston and Smith.[14] Similar procedures but with sublimation as the final step have been reported for $MgCl_2$.[15,16] The concentrations of residual impurities in the LiCl-containing melts, after the above purification, are less than $10^{-3}$ M.[12] The use of chlorine rather than anhydrous HCl in the purification of LiCl-containing mixtures has been described by Maricle and Hume,[17a] and also appears to lead to high-purity melts. Anhydrous chlorides have been prepared by refluxing the hydrated salt with thionyl chloride.[17b]

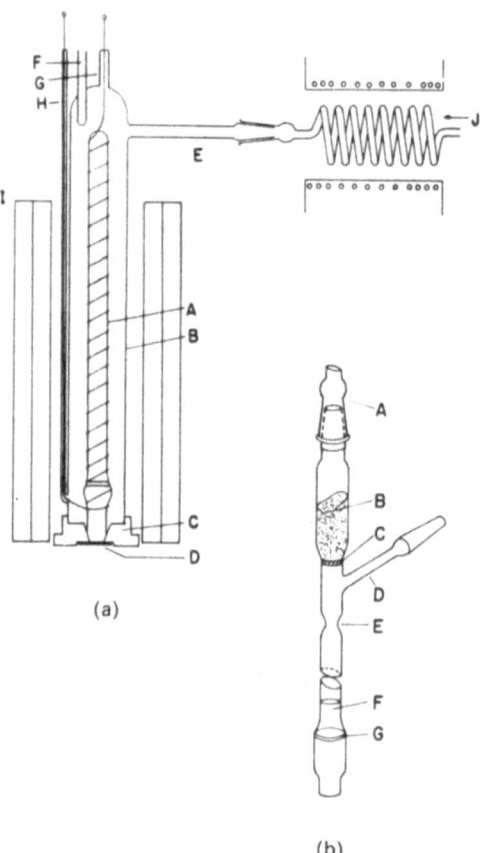

(a)

(b)

Fig. 1. Rensselaer Raman cell arrangement and molten salt filtration assembly. (a) Raman cell arrangement for high-temperature use: A, Raman cell (10–15 ml) wound with nichrome resistance heater; B, silica housing for Raman cell chamber; C, pedestal for Raman cell; D, removable glass window (optical flat); E, heated gas stream from preheater to Raman cell chamber; F, thermocouple well; G, conduit for heating coil; H, gas outlet tube; I, Dewar and filter solution vessel; J, gas inlet to preheater. (b) Molten salt filtration assembly: A, to vacuum manifold; B, salt sample (finely powdered); C, fritted Pyrex disk (fine porosity); D, sidearm to vacuum manifold; E, final seal constriction; $F_2$ molten salt after filtration; G, Raman cell window (optical flat).

Heavier alkali chloride melts can be quite adequately dried by slow heating under vacua; with chlorine or HCl treatments, the impurities are reduced from $10^{-3}$ M to undetectable levels.[17] Treatment of alkali chlorides with HCl in silica containers near 1000°C introduces dissolved silica impurities.[18] Water can be removed from lithium nitrate-containing melts by simple evacuation without apparent hydrolysis.[19]

Sublimation or distillation techniques are also excellent when applicable. Filtration for removal of small amounts of carbonaceous or other insoluble matter is accepted practice. For the preparation of certain salts, such as $ZnCl_2$ (i.e., exceedingly hydroscopic) in the anhydrous state, the reaction of anhydrous reactants has been used.[20] Other salts, e.g., $Hg(NO_3)_2 \cdot 2H_2O$, cannot be obtained as anhydrous single-salt melts due to inherent thermal instability. However, it has been reported[21,22] that when $Hg(NO_3)_2 \cdot 2H_2O$ is a solute in certain fused salts, water evolves spontaneously until the melt attains the anhydrous state. The preparation of anhydrous nitrate salts has been recently discussed by Addison.[23]

Tabulations of selected binary and ternary eutectic mixtures for the temperature range 100–500°C have been given by Janz et al.[3,4]

In Raman spectroscopy of melts, the chief problems* are the limitations imposed by the light source and intensity requirements. Where Pyrex glass or silica are suitable container materials, the Rensselaer arrangement, after Janz et al.,[24] as shown in Fig. 1(a) has been used successfully. This is designed for a Toronto arc[25] as excitation source. The cell of 14-ml capacity is wrapped with fine resistance wire and is supported in an outer quartz jacket. Temperature uniformity is controlled by a downward flow of preheated air or nitrogen through the annular space between the jacket and cell. With proper adjustments, a flat temperature zone at 500 ± 5°C along the entire cell length can be attained. This assembly is placed in a double-jacketed cylindrical Pyrex annulus having the inner compartment evacuated and in the outer, an appropriate filter solution. The "swollen" construction at the cell window is to ensure additional strength. An arrangement for molten salt filtration and cell filling is shown in Fig. 1(b). Similar cell assemblies heated only by a hot air stream have been described by Woodward et al.[26]

* For a discussion of infrared, and ultraviolet, and NMR methods, as well as Raman techniques, see R. A. Bailey and G. J. Janz.[4] See also D. W. James, "Vibrational Spectra of Molten Salts," in: B. R. Sundheim (ed.), *Fused Salts*, McGraw-Hill, New York, 1964.[1]

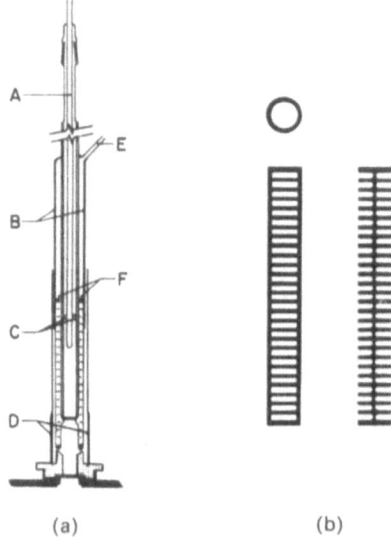

(a)                        (b)

Fig. 2. Chicago Raman cell arrangement
and graphite susceptor. (a) Arrangement
of cell in high-temperature assembly: A,
thermocouple well concentric in Raman
cell; B, light shield; C, melt level; D, light
shield; E, vacuum connection; F, graphite
susceptor. (b) Top and side views of
graphite susceptor. (Internal diameter,
3.12 cm; graphite rings are separated by
about 1.1-cm spacings.)

The Chicago assembly, after T. F. Young and co-workers,[27]
is shown in Fig. 2. This has been used at up to 1000°C in studies of
oxide melts, e.g., $B_2O_3$. Induction heating is achieved with graphite
susceptors of the design shown in Fig. 2(b) around the cell. Proper
spacing of the rings minimizes thermal gradients. Rather than allow
the melt to solidify in the cell, an auxiliary high-temperature furnace
is used to maintain the salt in the cell above its melting point once the
cell has been filled (i.e., a high-temperature thermostatted "storage
bath").

The Stuttgart arrangement, after Bues,[28] is shown schematically
in Fig. 3; it is of a radically different design. The molten salt is contained in an open crucible under a horseshoe-shaped mercury arc
source. The Raman scattering occurs within an inverted crucible

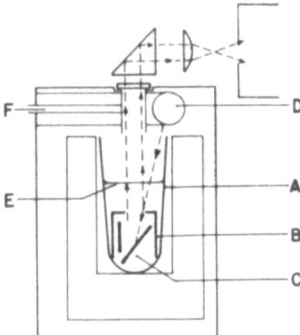

Fig. 3. Cross section of the Stuttgart Raman crucible cell arrangement. A, outer platinum crucible for sample containment; B, immersed platinum crucible with slit for Raman scattering; C, SiC crystals; D, mercury arc; E, melt level. The prism and crucible furnace can be adjusted by external screws for slit alignment.

totally immersed in the melt and having a slit on which the spectrometer is focused. A stream of preheated inert gas can be used to prevent deposition of volatile material on the window. This system has been used up to 850°C and can be used with highly reactive melts, since a wide range of materials may be used for the crucibles and the need for transparent cell walls does not enter in this arrangement.

Most Raman studies have been carried out with a mercury excitation source. This precludes the use of colored melts. Development of microwave-powered, radio-frequency, or laser-excitation sources[29-33] for long-wavelength Raman spectroscopy will extend the usefulness to colored systems.

## PRINCIPLES

Vibrational spectroscopy is used both as an analytical aid and for more precise quantitative studies of the structure and bonding of polyatomic species. The analytical method is based on the uniqueness of the spectrum of a molecule and on the relative insensitivity of force constants and vibrational frequencies as the environment about a group is changed.[34] Quantitative analyses rest on the concentration dependence of band intensities as given by the Beer–Lambert law in ideal cases.[34] This analytical approach is particularly useful in high-temperature spectroscopic problems where not infrequently the possibility of new species arises through inorganic reactions in the molten state.

Knowledge of chemical bonding and stereochemistry in inorganic melts can be advanced through the theory of molecular vibrations.[34,35] Group theoretical considerations show that it is often feasible to distinguish the arrangements of atoms in a polyatomic species from the number of vibrational bands observed in infrared and Raman spectra by reference to the point group, i.e., the totality of symmetry elements present in the species. In highly symmetric systems, some vibrations may be active in either infrared or Raman spectra, but not both. Degeneracy of vibrations may also limit the number of frequencies observed. In the least symmetric systems, degeneracy will not exist and all vibrations will be active in both infrared and Raman spectra. Tri-, tetra-, and pentaatomic species are most frequently encountered in ionic melts, and the vibrational spectroscopy of these

## TABLE I
### Selection Rules for $XY_2$, $XY_3$, and $XY_4$

| Species | Point group | Number of fundamentals of symmetry type and selection rules* |
|---|---|---|
| Y—X—Y | $D_{\infty h}$ | $\Sigma_g^+(R_p) + \Sigma_u^+(Ir) + \Pi_u(Ir)$ |
| X—Y—Y | $C_{\infty v}$ | $2\Sigma^+(R_p, Ir) + \Pi(R_d, Ir)$ |
| X on top with X and Y below (bent) | $C_{2v}$ | $2A_1(R_p, Ir) + B_2(R_d, Ir)$ |
| Y / X / Y Y (Planar) | $D_{3h}$ | $A_1'(R_p) + 2E'(R_d, Ir) + A_2''(Ir)$ |
|  | Removal of degeneracy in transition from $D_{3h}$ to $C_{2v}$ |  |
| Y / X / Y Y | $C_{2v}$ | $3A_1(R_p, Ir) + 2B_2(R_d, Ir) + B_1(R_d, Ir)$ |
| Y / X / Y Y (Pyramidal) | $C_{3v}$ | $2A_1(R_p, Ir) + 2E(R_d, Ir)$ |
| Y—X—Y (Tetrahedral) | $T_d$ | $A_1(R_p) + E(R_d) + 2F_2(R_d, Ir)$ |

* $R$ denotes Raman-active; $Ir$, infrared-active; $p$, polarized; $d$, depolarized.

is of particular interest in this review. The point groups, number of fundamentals of each type, and selection rules for various configurations of these species are given in Table I. The loss of degeneracy in the transition from a $D_{3h}$ system to a $C_{2v}$ system is also shown.

The selection rules based on "isolated" species hold best for molecules in the vapor phase. In condensed states, interactions between particles are often sufficient to cause splittings of degenerate frequencies and appearance of forbidden vibrational bands. Generally, the intensities of the bands allowed by such perturbations are less than those of active fundamentals. Due cognizance of this factor is important in assigning symmetry to condensed state systems based on the number of active frequencies.

In recent studies on silicate glasses,[36] it has been shown that appearance of a new frequency in infrared absorption spectra can be predicted theoretically from consideration of the gross structure of the melts and solids. The frequency and intensity of this band depend on the structure of the system as a whole. The Raman activity of the nitrate ion in molten thallous nitrate is such that two point-group symmetry classes, $D_{3h}$ and $C_{2v}$, appear possible on first inspection;[37] the former is ruled out when nearest-neighbor interactions are recognized in this ionic melt.[73]

The use of polarized radiation provides information about the symmetry type of a vibrational band. If polarized light is used in Raman spectroscopy, it can be shown[34] that the ratio of the intensity of light polarized perpendicular $I_\perp$ to the $XY$-plane to that polarized parallel $I_{\parallel}$ to this plane can have a maximum value of $\frac{6}{7}$. Here it is assumed that the direction of the incident radiation is along the $Z$-axis. In addition, antisymmetric vibrations will be depolarized, i.e., $\rho = \frac{6}{7}$, if they are permitted to occur at all. Only totally symmetric vibrations can be polarized, i.e., $\rho \leq \frac{6}{7}$. The state of polarization of Raman lines is indicated for the systems of interest in Table I.

Chantry and Plane[38] have used absolute Raman intensities to gain information about the nature of the bonding in ionic species in aqueous solution. The derivatives of molecular polarizability with respect to changes in bond length determine the intensities of Raman bands. It is recognized that meaningful results can be obtained only when the environment of a changing bond does not change; an excellent example is an isoelectronic series. Thus, in this manner, the $\pi$-bonding contribution in $NO_3^{-1}$, $CO_3^{-2}$ and $ClO_4^{-1}$, $PO_4^{-3}$ was investigated.[38] The results suggest that the $\pi$-bonding contributions in the $ClO_4^{-1}$ and

$NO_3^{-1}$ ions are considerably greater than those in $CO_3^{-2}$ and $PO_4^{-3}$.

Normal coordinate vibrational analyses[35] may provide information about chemical bonds. In this method, force constants are calculated by an iterative procedure using an assumed form of the potential function. The set of force constants obtained is usually not unique, but comparison of force constants for systems of the same symmetry gives an indication about the nature of the bonding in polyatomic systems. An illustration is the vibrational analysis[34e] of the planar $XY_3$ group using the Urey–Bradley potential field for six inorganic species ($BF_3$, $BCl_3$, $BBr_3$, $NO_3^{-1}$, $CO_3^{-2}$, $BO_3^{-3}$). The repulsion force constants in these species are found to correlate directly with the separation of the nonbonded atoms of the species $XY_3$. The deviation of the $NO_3^{-1}$ ion from this correlation, due to the greatly shortened length of the N–O bonds, is understood by the enhanced localization of the positive and negative charge densities in the resonance hybrids relative to the species considered.[34f] This type of analysis is also required as a precursor to the interpretation of intensity measurements such as those made by Chantry and Plane.[38]

Significant contributions to the structural problems in inorganic molten salt systems may be foreseen with the advance of such spectroscopic methods into the area of high-temperature chemistry.

## MELT SPECTRA

### Single-Salt Melts

Melts from pure salts may be classified broadly in two groups—those from salts recognized as essentially ionic in the crystalline state (e.g., NaCl), and those from salts having a pronounced covalent component of bonding in the solid state (e.g., $ZnCl_2$ and $HgCl_2$). The former liquids consist predominantly of ions corresponding to the simple constituents of the solid-state lattice; in the latter, the melt may retain "fragments" of the original solid-state species as molecular entities and complex ions, as well as the more simple anionic and cationic components. Studies with salts in the first group have been largely directed to investigations of the nearest-neighbor interactions to contribute to the concept of liquid-state theory, e.g., the quasi-crystalline model. Salts from the second group are more suited for investigations of problems of chemical bonding and coordination

## TABLE II
### Vibrational Spectra of Molten Nitrates*

| | Frequency ($cm^{-1}$) | | | | Lattice mode | Reference |
|---|---|---|---|---|---|---|
| | $\nu_1$ | $\nu_2$ | $\nu_3$ | $\nu_4$ | | |
| $LiNO_3$ | 1067 | 821 | 1375, 1460 | 726 | 343 | 43 |
| $NaNO_3$ | 1053 | 827 | 1412 | 722 | 238 | 43 |
| $KNO_3$ | 1048 | 829 | 1391 | 718 | < 220 | 43 |
| $RbNO_3$ | 1046 | — | 1372 | 713 | < 220 | 43 |
| $CsNO_3$ | 1043 | — | 1356 | 708 | — | 43 |
| $AgNO_3$ | 1043 | 800 | 1310, 1395 | 727 | — | 43 |
| $TlNO_3$ | 1036 | 813 | 1328, 1383 | 708 | — | 43 |

* $D_{3h}$ point-group symmetry.

interactions and the inorganic chemistry of the molten state. The nitrate ion has been used as a spectroscopic "detector" in ionic melts in the former studies; the halides of the metals of Groups II and III appear best suited for the latter type of investigations.

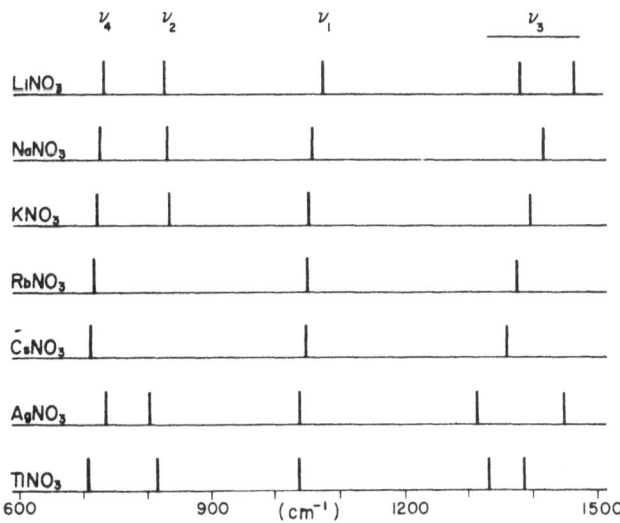

Fig. 4. Vibrational frequencies of the nitrate ion in molten nitrates.

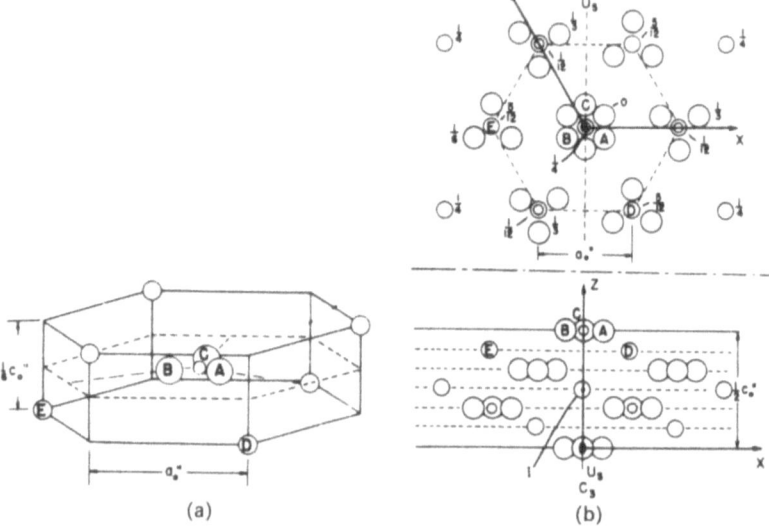

Fig. 5. (a) The arrangement of cations around the $NO_3^-$ ion in rhombohedral crystalline $NaNO_3$. (b) Crystal symmetry of $NaNO_3$. The symmetry is $D_{3d}^6$ (rhombohedral). The largest circles represent oxygen atoms, the intermediate circles sodium atoms, and the smallest circles nitrogen atoms. A projection to the $XZ$ plane is also shown. $i, C_3, U_3$, and $\Delta$ have their usual meanings; $a_0''$ and $c_0''$ are the units for the pseudohexagonal cell: $a_0'' = 5.06$ Å, $c_0'' = 16.81$ Å.

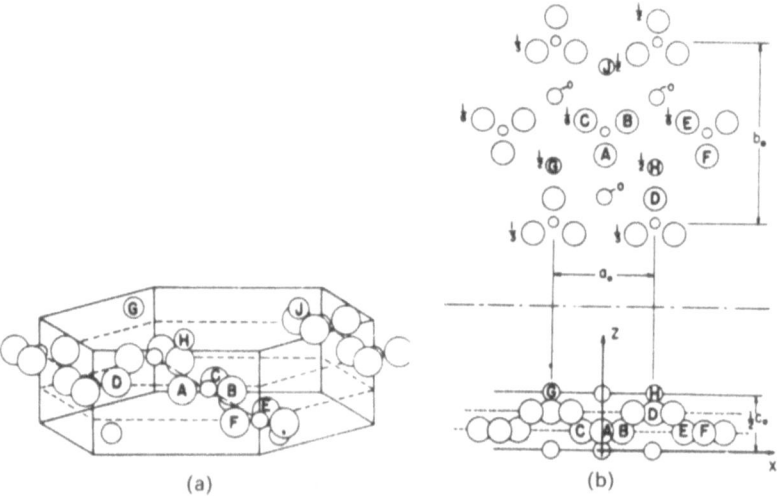

Fig. 6. (a) The arrangement of cations in the orthorhombic $KNO_3$ crystal. (b) Crystal structural of $KNO_3$. The symmetry is $V_h^{16}$ (orthorhombic). The largest circles represent oxygen atoms, the intermediate circles potassium atoms, and the smallest circles nitrogen atoms. A projection to the $XZ$ plane is shown. $a_0$, $b_0$, and $c_0$ are the units of the cell: $a_0 = 5.41$ Å, $b_0 = 9.16$ Å, $c_0 = 6.43$ Å.

**Nitrates.** The vibrational spectra of nitrate ions have been extensively investigated in aqueous solution[39] and in crystals.[40] The past decade has seen the extension of such studies to nitrates in the molten state.[41] Table II lists data for the melts of the pure alkali metal nitrates and the silver and thallous salts. The bar graph representation of Fig. 4 shows the spectral changes as the cationic species is varied in these molten nitrates. Inspection shows three salient features: (1) loss of degeneracy of the $E'$ modes in certain nitrates; (2) the observation of a low frequency, i.e., a band in the range $300–200 \text{ cm}^{-1}$; and (3) correlation of the shift of the symmetrical stretching frequency with change in the cationic species in the molten nitrates.

For the alkali metal nitrates, the degeneracy of the $v_3$ fundamental mode is not lost except for $LiNO_3$ (Table II). In this case, both infrared and Raman spectral studies confirm the presence of two bands, one centered at about $1375 \text{ cm}^{-1}$ and the other at about $1455 \text{ cm}^{-1}$. Molten thallous nitrate[41e] has bands at 1328 and $1383 \text{ cm}^{-1}$, while molten silver nitrate has bands at 1310 and $1395 \text{ cm}^{-1}$. More recently Walrafen and Irish[41g] have reinvestigated the Raman spectrum of molten $AgNO_3$. An analysis of the band intensities and contours revealed the presence of nine Raman frequencies. These new and significant features are attributed, in part, to vibrational perturbations.

Absorption bands in the infrared spectrum of molten lithium, sodium, potassium, and silver nitrates have been observed at 343, 238, and $< 220 \text{ cm}^{-1}$, respectively.[41b] These bands have been assigned[41b] to "lattice-like" vibrations in the melt, presumably not unlike the lattice vibrations of a crystalline solid. The existence of such modes does not necessarily imply long-range order in the molten salt, but may be attributed to the pronounced short-range, strong cation–anion interactions between the nitrate ion and its nearest neighbors (see, for example, the work of Wait and Ward[41h]). Schroeder, Weir, and Lippincott[40a] have shown that a librational mode in solid nitrates might be expected in the vicinity of $20–40 \text{ cm}^{-1}$. Librational modes are normally weak and occur in a portion of the spectrum difficult to observe by the Raman and conventional infrared techniques; the methods of far-infrared spectroscopy are needed to advance studies in this region.

The symmetrical stretching frequency $v_1$ shows a definite correlation with the properties of the cation. A systematic variation in $v_1$ is observed for the following series of molten nitrates: $LiNO_3$ $(1067 \text{ cm}^{-1})$; $NaNO_3$ $(1053 \text{ cm}^{-1})$; $KNO_3$ $(1048 \text{ cm}^{-1})$; $RbNO_3$

($1046 \, \text{cm}^{-1}$); and $CsNO_3$($1043 \, \text{cm}^{-1}$). Various empirical correlations with ion size and polarizability have been advanced.[41,73]

In studies of both the electronic[42] and vibrational[43] spectra of alkali nitrates, the nitrate ion may act as a sensitive detector of coulombic interactions and packing effects. The bonding of the nitrate ion consists of a framework formed by overlap of $sp^2$ orbitals on the nitrogen with $p$ orbitals on the oxygen and $\pi$-bonds in which there is a large measure of delocalization. The model of the nitrate ion proposed for the molten nitrates[43] is a planar disklike species (4.62-Å diameter, 2.2-Å thickness) evolved by rotation of the ion about its $C_3$-axis. The volume, 24.8 Å$^3$, corresponds exactly to that of a spherical $Cl^-$ ion. Spectroscopically, the nitrate ion in its "free" state belongs to the point group $D_{3h}$ and, accordingly, should have one fundamental active in the Raman effect ($A_1'$), one in the infrared spectrum ($A_2''$), and two active in both the infrared and Raman spectra ($E'$) as listed in Table I. If covalently bonded as in methyl nitrate or certain nitrate coordination compounds, the nitrate obtains the $C_{2v}$ group symmetry. The transitions and correlation between these two point groups are summarized in Table I.

In solid sodium nitrate, no splitting of $v_3$ is observed,[41] while in solid potassium nitrate, this degeneracy is removed.[39,41a,f] Crystalline sodium nitrate belongs to the rhombohedral space group $D_{3d}^6$ (calcite type), while potassium nitrate is in the orthorhomic space group $V_h^{16}$ (aragonite). A recent infrared investigation of crystalline modifications of potassium nitrate confirms the important role of site symmetry on the activity and loss of degeneracy of vibrational transitions in crystals.[41f] As seen in Fig. 5, the nitrate ion in the former case lies on a threefold symmetry axis. Accordingly, the spectrum should not lose its degeneracy. In the aragonite structure of $KNO_3$, illustrated in Fig. 6, the nitrate ion is no longer on a threefold crystal axis; it is subjected to an unsymmetrical environment, and loss of the degeneracy for $v_3$ of the isolated ion is a consequence. Extension of this concept to molten nitrates follows. Four "contact" sites seem most probable: (1) above the plane of the ion and directly on the threefold axis or a "top" site; (2) along the N–O bond or a "corner" site; (3) bisecting the O–N–O angle or a "crook" site; and (4) a position intermediate to (1) and (3) or a "roll-on" site. These are shown in Fig. 7 as A, B, C, and D, respectively. Both of the B and C sites give rise to an unsymmetrical field which might remove the degeneracy in the nitrate ion. Support for this viewpoint has recently been advanced

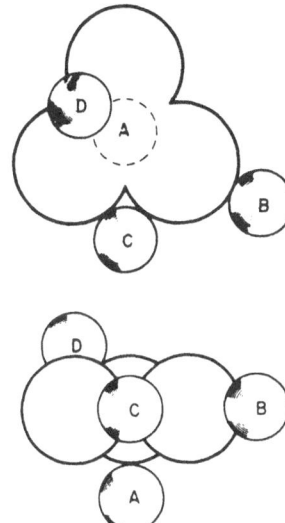

Fig. 7. Top and side views of nonequivalent cation sites around the nitrate ion. The four positions are: A, the "top" site; B, the "corner" site; C, the "crook" site; and D, the "roll-on" site.

from analyses of X-ray and neutron diffraction data and physico-chemical properties for molten nitrates[44] and molten carbonates.[45] For nitrates in the molten state near the melting temperatures, the radial distribution analyses of the diffraction data confirm that, for example, in $NaNO_3$ the most probable positions of $Na^+$ ions around $NO_3^-$ ions are the "corner sites" B (Fig. 7) and intermediate to positions A and C, i.e., D (the "roll-on" sites). In molten carbonates it has been shown similarly that in passing from lithium to potassium salts the cations assume average positions about these planar disklike anions which are increasingly close to the corner sites, i.e., position B (Fig. 7). From another viewpoint, it is clear that such interactions, sufficiently enhanced, will be understood as symmetry changes (nitrate ion) from point group $D_{3h}$ to $C_{3v}$ or $C_{2v}$.

In molten nitrates the presence of ion pairs of $C_{2v}$ or $C_s$ symmetry (or possibly, but less likely, $C_{2v}$ or $C_s$ symmetry, involving a covalent N–O–metal link) would produce spectra different from the $D_{3h}$ species. Interpretation of the spectral features can be advanced using vibrational theory.

Normal coordinate vibrational analyses for the nitrate ion have been carried out by several investigators;[34e,f;40k;41a,d] most of these have assumed "free" nitrate ions of symmetry $D_{3h}$, thereby predicting a single frequency for each of the fundamentals. More recently, the spectra of $AgNO_3$, $TINO_3$, and the alkali metal nitrates have been analyzed using an ion-pair model.[41h,73]

The observed spectra can be satisfactorily accounted for by either the $C_s$ or $C_{2v}$ point-group symmetry of the anion–cation pair species. The simple pair model serves to explain many features of the infrared and Raman spectra but does not preclude more extensive coordination or site-symmetry effects as shown by Walrafen[46] and Devlin.[47] Two low frequencies ($\sim 190$–150 and 93–66 cm$^{-1}$) are predicted; these would normally be assigned to metal–oxygen interactions in the anion–cation species. Far-infrared spectral studies for molten $AgNO_3$, as required to establish this assignment, are an outstanding need. The splitting of the degenerate fundamental and the appearance in the Raman spectra of the forbidden fundamental ($v_3$, $D_{3h}$) have been proposed as the spectroscopically based operational critiques for ion

## TABLE III

### Vibrational Spectra of Molten Hydroxides, Chlorates, Perchlorates, Sulfates, and Bisulfates

| Compound | Frequency (cm$^{-1}$) | | | | Lattice mode | Reference |
|---|---|---|---|---|---|---|
| | $v_1$ | $v_2$ | $v_3$ | $v_4$ | | |
| Hydroxides: | | | | | | |
| LiOH | 3652 | | | | 440 | 49 |
| NaOH | 3295 | | | | 265 | 49 |
| KOH | 3300 | | | | 233 | 49 |
| Chlorates: | | | | | | |
| LiClO$_3$ | 938 | 620 | 1018, 977 | 478 | 338 | 50 |
| NaClO$_3$ | 932 | 611 | 988 | 482 | 200 | 50 |
| KClO$_3$ | 931 | 603 | 981 | 489 | 200 | 50 |
| AgClO$_3$ | 895 | 595 | 969,931 | 477 | 200 | 50 |
| Perchlorates: | | | | | | |
| LiClO$_4$ | 948 | 470 | 1139 | 627 | 315, 260 | 50 |
| Sulfates: | | | | | | |
| Li$_2$SO$_4$ | 987 | 460 | | 625 | | 51 |
| Na$_2$SO$_4$ | 972 | 460 | | 625 | | |
| Bisulfates: | | | | | | |
| KHSO$_4$ | 1230 | 1050 | 840 | $\dfrac{v_{4,5,8} \; v_{6,9} \quad v_7}{590 \;\; 420 \;\; 1390\text{–}1360}$ | | 52 |

association (ion pairing) in aqueous solution for a variety of inorganic nitrates.[34e,48a] It would appear that this may equally well be applied to the molten state for these salts, with the difference that in the inorganic melts the ion–ion interaction must be contact ion pairs as distinct from the solvent-separated ion pairs of aqueous and non-aqueous solution at ambient temperatures.[48b]

**Hydroxides, Chlorates, Perchlorates, Sulfates, and Bisulfates.** Molten alkali hydroxides,[49a] chlorates,[50] perchlorates,[50] sulfates,[51a] and bisulfates[51b] have been studied. The data are given in Table III. In the infrared studies, the hyroxides, chlorates, and perchlorates all show a low-lying vibrational frequency similar to that observed in the case of the nitrate melts. It is significant that the bands at 440, 265, and 233 cm$^{-1}$ in lithium, sodium, and potassium hydroxide, respectively, are the most intense absorptions in the hydroxide spectra. The O–H stretching frequency occurs at $\sim 3500$ cm$^{-1}$ in these melts, but is weak and split by 350–400 cm$^{-1}$. This splitting has been suggested as due to the P and R branches of a freely rotating (O–H)$^-$ ion or alternately to hydrogen-bonding effects. Consideration of the solid-state spectrum of lithium hydroxide confirms the coupling of the OH$^-$ ions.[49b]

The chlorate ion in aqueous solution has been studied by Chen and co-workers[52] and in the solid by Rocchiccioli.[53] The assignments were made on the basis of $C_{3v}$ symmetry, i.e., all three oxygens equivalent. The investigation of molten chlorates[50] seems to verify that $C_{3v}$ symmetry is retained in the melt, if allowance is made for splitting of the degenerate vibrations due to environmental effects.

The totally symmetric vibrations ($v_1, v_2$) show a decrease in the series lithium, potassium, silver, the frequencies being (938, 620), (931, 603), and (895, 595) in these cases, respectively. The higher of the two doubly degenerate fundamentals $v_3$ shows a similar decrease; the lower fundamental $v_4$ shows a definite increase, e.g., Li$^+$ (478), K$^+$ (489), and in the molten silver salt loses its degeneracy (477, 440).

In molten LiClO$_4$, the tetrahedral ($T_d$ point group) symmetry of the ClO$_4^-$ ion is apparently retained. The spectrum consists of two strong bands at 1139 and 627 cm$^{-1}$ and two weaker bands at 948 and 470 cm$^{-1}$. The latter two are probably the infrared-inactive fundamentals; the appearance of these forbidden frequencies as weak infrared lines is undoubtedly due to condensed-state environmental factors.

Raman spectral studies of potassium bisulfate have been reported by Walrafen, Irish, and Young[51b] over the temperature range from 300 to 700°C. The spectra indicate that reactions that form the $S_2O_7^{-2}$ and $SO_4^{-2}$ ions occur. Assignments of the vibrations of $HSO_4^-$, $S_2O_7^{-2}$, $SO_4^{-2}$ are made and correlated with results in aqueous solution. It appears that the $HSO_4^-$ ion belongs to a $C_s$ group.

The Raman spectra of molten and aqueous lithium and sodium sulfates have been investigated by Walrafen.[51a] The tetrahedral ($T_d$) symmetry of the sulfate ion is retained in both the aqueous solutions and the melts. In the spectra of the molten sulfates, shifts in the values of the totally symmetric $SO_4^{-2}$ stretching frequencies are noted with variations of the cationic species, much as in the spectra of nitrates as single-salt melts. Intensity ratios for the symmetrical and asymmetrical stretching frequencies are reported and examined relative to specific anion–cation interactions in molten sulfates.

**Molten Halides.** The halides of metals of Group II (predominantly covalent in the solid state[54] and the vapor state[55]) remain largely unexplored in the molten state by vibrational spectroscopy. In the solid-state layer-type structure for $CdCl_2$, illustrated in Fig. 8, the triatomic linear species can be recognized, and the unsymmetrical cationic environment is evident. Selection rules for "free" molecules of this type are given in Table I. Spectral data for these compounds in the molten state are summarized in Table IVa.

The Raman spectrum of $ZnCl_2$ as a single-salt melt has been the subject of a series of studies.[54,56] The intense band at $305\,cm^{-1}$ is

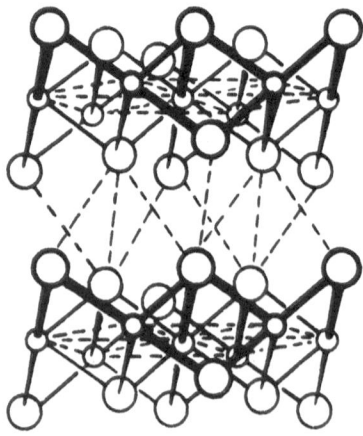

Fig. 8. The layer-type structure in solid $CdCl_2$.

## TABLE IVa
### Vibrational Spectra of Molten Halides

| | Frequency (cm$^{-1}$) | | | Reference |
|---|---|---|---|---|
| | $\nu_1$ | $\nu_2$ | $\nu_3$ | |
| $ZnCl_2$ | 305 | | | 56b,c |
| $ZnBr_2$ | 155 | | | 67b |
| $CdCl_2$ | 212 | | | 67b |
| $MgCl_2$ | 270 | 214 | 143 | 57 |
| $HgCl_2$ | 313 | 376 | 100 | 59 |
| $HgBr_2$ | 195 | 271 | 90 | 59 |
| HgBrCl* | 236 ($\nu_1$) | 335 ($\nu_3$) | 111 ($\nu_2$) | 59 |

* $C_{xv}$ symmetry. $\nu_1$ and $\nu_3$ are stretches; $\nu_2$ is a bending mode.

assigned[56b,c] to the symmetric stretching frequency $\nu_1$ of the $ZnCl_2$ triatomic species. Two studies have reported detailed spectra for the single-salt melt, i.e., at 95, 110, 230, 266, and 375–400 cm$^{-1}$,[56b] and 75, 226, 250, and 360 cm$^{-1}$,[56c] in addition to the 305 cm$^{-1}$ frequency. The intensity of the band at 230 cm$^{-1}$ decreases with the addition of water;[56b] this band is also observed in the spectra of aqueous solutions more concentrated than 10 moles/liter.[56d] The various species and the vibrational assignments are in Table IVb. In molten $ZnCl_2$ it appears that a polymer formulated as $(ZnCl_2)_n$ is a dominant species, with the $ZnCl_4^{-2}$ as a recurring unit in which each tetrahedron is linked by

## TABLE IVb
### Characteristic Vibrational Bands in Halide Melts

| | Species | Frequency (cm$^{-1}$) | Reference |
|---|---|---|---|
| | $(ZnCl_2)_n$ polymer | 75, 226, 250, 360 | 56c |
| $ZnCl_2$ | $ZnCl_2$ polymer | 95, 230 | 56b |
| | $ZnCl_n^{2-n}$ | 110, 266 | 56b |
| | $MgCl_6^{-4}$ | 143, 214, 270 | 57 |
| $MgCl_2$ | $MgCl_3^-$ | 158, 208, 251, 450 | 57 |
| | $MgCl_4^{-2}$ | 232 | 57 |
| $CdCl_2$ | $CdCl_3^-$ | 177, 211, 245, 257 | 70 |
| $AlCl_3$–NaCl | $AlCl_4^-$ | 145, 183, 349, 580 | 67g |
| $AlCl_3$–KCl | $AlCl_4^-$ | 136, 155, 168, 182, 196, 353, 505, 562, 597 | 67g |
| $HgCl_2$ | $HgCl_3^-$ | 210, 282, 287 | 59 |
| | $HgCl_4^{-2}$ | 180, 192, 267, 276 | 59 |

a common chloride to neighboring tetrahedra as in the crystal. This polymeric association is diminished with increasing temperature. In addition to the polymer, the existence of discrete molecules ($ZnCl_2$ or $ZnCl_n^{2-n}$) is supported.

The spectrum of $MgCl_2$[57] in the molten state also indicates bridging structure. However, the recurring species has been interpreted as $MgCl_6^{-2}$ ions. Bands at 269, 241, 214, and 142 cm$^{-1}$ are assigned to the octahedral complex.

The vibrational spectra of molten $HgCl_2$ and $HgBr_2$[58-60] are understood if the melts are composed of linear triatomic species $D_{\infty h}$ symmetry. The finite electrical conductance of these melts has been attributed[58] to the dissociation of the molecular solvent species with the formation of species $MX^+$, $MX_3^-$, $MX_4^{-2}$ in amounts less than 0.01%. In this concentration range, the ionic species are below the level necessary for spectroscopic detection. The symmetric stretching frequency $v_1$ is broad in the Raman spectrum of the melts (e.g., $HgCl_2$; $v_1$, 313 cm$^{-1}$; line width, $\sim 80$ cm$^{-1}$). The band broadening has been attributed to the heteropolar nature of the M–X bond and the possibility of perturbation of the halide orbitals by molecular rotational interactions.[59] The appearance of the Raman forbidden frequency $v_2$ is additional support for this viewpoint.

An insight into the bonding processes in mercuric halides has been gained from ambient temperature studies of the "solvent-shift" spectra[61,62] (totally symmetric stretching vibration) and the spectra of the vapor-phase species.[55] The bonding orbital of the mercury is apparently not due to purely covalent $sp$ hybridization, but rather also partly formed by the interaction of a repulsive lower state with an attractive $sp$ upper state. The degree of ionicity of the Hg–Cl bond has been estimated as 23% from studies of the "solvent-shift" spectra. A comparison of the interaction force constant values for the gaseous, solid, and molten states of $HgCl_2$ [$HgCl_2$, $k_{12} = 0.037$ dyne/cm (gas),[55] $-0.06$ dyne/cm (solid), and $-0.07$ dyne/cm (molten)[59]] shows that this parameter is a sensitive index for environmental effects (interparticulate force fields). High-temperature infrared studies for the molten mercuric halides remain an outstanding need.

The Raman studies of molten gallium dichloride and dibromide by Woodward and co-workers[63a,b] provide yet a further illustration of the contributions to a better understanding of the inorganic molten state. Possible species suggested were $X_2Ga$–$GaX_2$, $GaX_4^-$, and related ionic entities, such as $(GaX_2$–$GaX)^+$. The results, summarized in

## TABLE V
### Raman Spectra of Gallium Dichloride and Dibromide*

| Assignment ($T_d$) | Aqueous solution | | Molten | |
|---|---|---|---|---|
| | $GaCl_4^-$ | $GaBr_4^-$ | Dichloride | Dibromide |
| $v_2(E)$ | 114 | 71 | 115 s, dp, sh | 79 s, dp, sh |
| $v_4(F_2)$ | 148 | 102 | 153 w, dp, dif | 107 s, dp, sh |
| $v_1(A_1)$ | 346 | 210 | 356 vs, p, sh | 209 vs, p, sh |
| $v_3(F_2)$ | 386 | 278 | 380 vw, dp, dif | 288 w, dp, dif |

* Taken from the work of Woodward *et al.*[63a,b] and Nixon and Plane.[63c]
Notation: s, strong; dp, depolarized; p, polarized; sh, sharp; dif, diffuse;
m, medium; w, weak; vw, very weak. Values are in units of $cm^{-1}$.

Table V together with the data for the species $GaX_4^-$ in aqueous solution, leave no doubt as to the nature of the ionic species in the molten state of these halides. Even after long exposures, no Raman activity characteristic of $GaCl_3$ is observed in molten gallium dichloride. Evidence for nearest-neighbor ionic interactions in these melts is seen in the fact that the intensity of the triply degenerate fundamentals for $GaX_4^-$, i.e., $v_3$ was weaker and more diffuse in the molten salts relative to the corresponding salts in aqueous solutions.[63c]

## TABLE VI
### Raman Spectra of Thiocyanates

| Salt | Temperature (°C) | Frequency ($cm^{-1}$) | | |
|---|---|---|---|---|
| | | $v_1$(p, vs) | $v_2$(dp, w) | $v_3$(p, m) |
| Molten Thiocyanates: * | | | | |
| LiCNS | 185 | 2083 | 499 | 764 |
| NaCNS | 320 | 2074 | 490 | 745 |
| KCNS | 190 | 2068 | 478 | 745 |
| Aqueous Thiocyanates: | | | | |
| LiCNS | | 2072 ± 2† | 496 ± 1† | 745 ± 4 |
| NaCNS | | 2076 ± 2 | 488 ± 2 | 744 ± 3 |
| KCNS | | 2074 ± 1 | 488 ± 2 | 746 ± 2 |
| RbCNS | | 2074 ± 2 | 488 ± 2 | 747 ± 2 |
| CsCNS | | 2073 ± 1 | 485 ± 3 | 747 ± 2 |

* Taken from the work of Baddiel and Janz.[63d]
† Average for concentration range, 1.0–13.6 m.

**Molten Thiocyanates.** The Raman spectra assignments[63d] for molten lithium, sodium, and potassium thiocyanate (Table VI) confirm that the thiocyanate ion in the molten state of these salts is a linear triatomic species. The alkali metal thiocyanates appear to form highly ionic and completely dissociated melts on fusion, in which the thiocyanate ions are "kinetically free" entities. Systematic changes in the $C \equiv N$ stretching frequency $v_1$ are noted as the cation size is changed.

Examination of the Raman results for the corresponding salts in aqueous solutions reveals that there is no change in vibrational frequency with dilution, and, with the exception of aqueous lithium thiocyanate for which a slight shift to lower frequencies is noted (Table VI), that there is little difference between the melt frequencies and those of the solutions. The "levelling" effect of ionic solvation in the aqueous solutions apparently swamps out any dependence of the frequency $v_1$ on the cationic field (*cf.*, molten salt spectra, Table VI).

In the most concentrated solution (13.6 m), the solute–solvent ratio is 1:4 and the mole fraction of solute is 0.2. The results gained from these very concentrated solutions confirm that the removal of last traces of solvent makes little difference to the vibrational spectrum of the pure salt compared with that of the salt as a single-salt melt. It follows, for systems where crystal hydrates are not formed, that a good estimate of the melt spectrum and structure can be gained from studies of ultraconcentrated solutions at ambient temperatures.

## MIXTURES

Spectroscopic investigations of molten two-component systems are more limited in number than those for pure molten salts. These studies may be broadly divided into three types: (1) melts of two or more compounds, each of which is predominantly ionic; (2) melts formed from one ionic compound as a solute in a solvent of a covalent inorganic substance; and (3) melts formed from two or more covalent compounds. There follows a survey of the systems most recently examined with specific reference to chemical interactions and the formation of new species.

In the first class of melts, the predominant spectroscopic interest has been the investigation of changes in vibrational spectra with composition. Bues[64] examined the spectrum of an equimolar melt of

## TABLE VII
### Vibrational Spectra of Molten Nitrate Mixtures

| System | Frequency (cm$^{-1}$) | | | | | |
|---|---|---|---|---|---|---|
| | $\nu_1$ | $\nu_2$ | $\nu_3$ | $\nu_4$ | $2\nu_2$ | Reference |
| AgNO$_3$–KNO$_3$ (1:1) | 1043 | 817 | 1312, 1425 | 720 | 1633 | 64 |
| NaNO$_3$–5% LiNO$_3$ | 1054 | — | 1308–1472 | 723 | 1665 | 65 |
| NaNO$_3$–10% LiNO$_3$ | 1057 | — | 1319–1484 | 724 | 1656 | 65 |
| NaNO$_3$–30% LiNO$_3$ | 1061 | — | 1319–1493 | 723 | 1650 | 65 |
| NaNO$_3$–60% LiNO$_3$ | 1061 | — | 1305–1398 | 724 | 1646 | 65 |
| | | | 1398–1495 | | | |
| NaNO$_3$–50% KNO$_3$ | 1050 | — | 1396–1492 | 718 | 1655 | 65 |
| Ca(NO$_3$)$_2$–60% KNO$_3$ | 1049 | — | 1300–1361 | 713 | 1646 | 65 |
| | | | 1407–1500 | 733 | | |

AgNO$_3$–KNO$_3$. The vibrational frequencies were intermediate between those for pure AgNO$_3$ and pure KNO$_3$, and the spectrum retained the same general features, i.e., splitting of degeneracy and intensity ratios, as for pure AgNO$_3$. Mixtures of sodium nitrate and lithium nitrate in the concentration range from 5 to 60% LiNO$_3$ have been studied.[65] A linear variation in the symmetric stretching frequency $\nu_1$ is noted as composition is changed. For KNO$_3$–Ca(NO$_3$)$_2$ mixtures, glassy systems with complex spectra are reported.[66] The removal of degeneracy of both $\nu_3$ and $\nu_4$ in this mixture has also been observed.[65] A summary of the preceding data is found in Table VII.

Melts from mixtures of ionic and covalent halides have received more attention.[67] A striking feature is the appearance of new bands as the ionic halide is added to the covalent melt. In the case of molten mercuric chloride–potassium (or ammonium) chloride systems, it has been shown[67a] that the species HgCl$_3^-$ and HgCl$_4^{-2}$ are formed. Studies for mercuric halide–alkali halide in aqueous and organic solvent systems confirm the formation of these complex ions.[68] The variation of the symmetric stretching frequency $\nu_1$ for each of the three species HgCl$_2$, HgCl$_3^-$, and HgCl$_4^{-2}$ in the molten salt mixtures as excess Cl$^-$ is added as an additional ligand is given in Fig. 9.

Normal coordinate vibrational analyses for HgCl$_2$,[59] HgCl$_3^-$,[67a] and HgCl$_4^{-2}$,[67a] have been reported for the Urey–Bradley potential field. The symmetrical stretching force constant $k_1$ decreases in the order HgCl$_2$ (2.11), HgCl$_3^-$ (1.27), and HgCl$_4^{-2}$ (0.82), and this undoubtedly reflects the changes in the nature of the bonding processes in the Hg–Cl bond in the inorganic molten state as excess $Cl^-$ is added. If the ionicity of the Hg–Cl bond is taken as $\sim 23\%$ in molten $HgCl_2$, the ratios of the values of $k_1$ predict that the ionicity of this bonding is $\sim 50$ and $70\%$, respectively, in $HgCl_3^-$ and $HgCl_4^{-2}$.

The compositions of $HgCl_2$–$TlNO_3$ mixtures have been investigated[67f] by Raman spectroscopy over the complete range of concentrations from one single-salt melt to the other. Independently, cryoscopic studies in thallous nitrate, with $HgCl_2$ and $HgBr_2$ as solutes, are understood if the mercuric halide solutes dissolve as simple molecular species in molten thallous nitrate.[69] The Raman data[67f] show clearly that the most intense Raman lines in the mixtures of these two salts over the entire composition range are the symmetrical stretching frequencies for $NO_3^-$ ($1036\,cm^{-1}$, pure molten thallous nitrate) and for $HgCl_2$ ($313\,cm^{-1}$, pure molten $HgCl_2$). While species such as

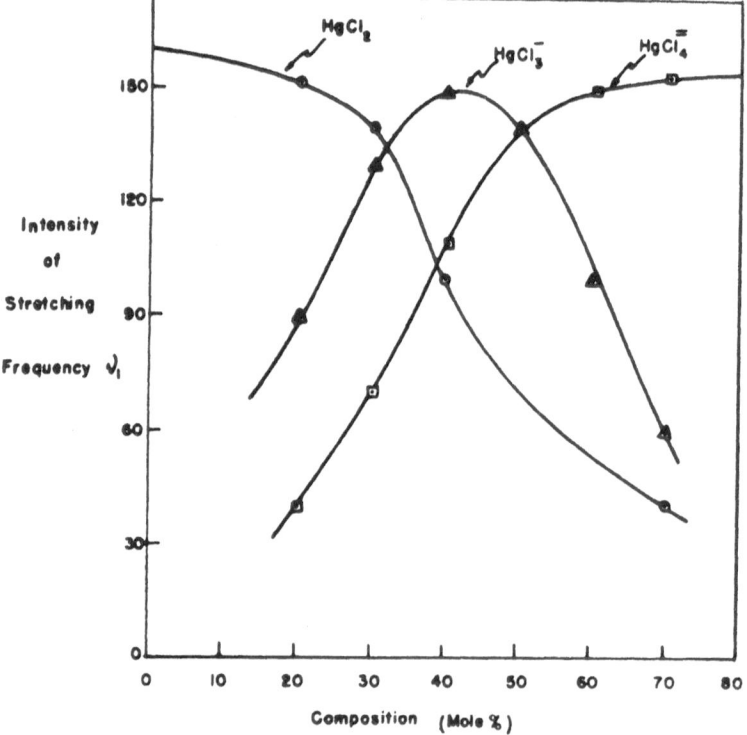

Fig. 9.  Variation of intensity of $v_1$ with composition (excess $Cl^-$ added as ligand).
⊙ $HgCl_2$.  △ $HgCl_3^-$.  ☐ $HgCl_4^{-2}$.

$HgCl_2(NO_3)^-$ and $HgCl_2(NO_3)_2^{-2}$ are not ruled out, the interactions are most likely limited to solvation-type forces in the thallous nitrate–mercuric halide molten salt mixtures.

Studies of molten mixtures of two covalent compounds appear virtually nonexistent. A report of the spectrum of $HgBrCl^{67}$ indicates that this compound disproportionates in the molten state into $HgCl_2$ and $HgBr_2$.

Molten complex salts of Cd(I) with aluminum chloride have been investigated using Raman spectroscopy.[70] The presence of $Cd_2^{+2}$ was demonstrated by a strong band at 183 cm$^{-1}$. The presence of complex ions in molten mixtures of Cd(II) halide–alkali metal halide has been confirmed by Raman studies and various physico-chemical and electro-chemical studies.[71] The stoichiometry and stereochemistry of these complex ions are current problems. Ions such as $CdCl_3^-$ (both as planar and pyramidal species) and $CdCl_4^{-2}$ (tetrahedral symmetry) have been advanced as the predominant species in 50:50 mol.% $CdCl_2$–KCl molten mixtures. The thermodynamics of these molten salt mixtures has been qualitatively explained[72] with a model in which $CdCl_4^{-2}$ is the important complex ion, and species such as $CdCl_3^-$ and $CdCl_6^{-4}$ are present in only minor concentrations. Recent investigations of the Raman spectra for molten mixtures of $CdCl_2$–KCl (33:66 mol.% KCl) have shown[71d] that the vibrational assignment, intensities, and polarization data are in accord with a system where the dominant complex ion in this concentration range is $CdCl_3^-$ and that symmetry of this species is pyramidal rather than planar.

In the mixtures $MgCl_2$–KCl,[57] $AlCl_3$–KCl,[67g] $AlCl_3$–NaCl,[67g] $ZnCl_2$–KCl,[56c] the predominant effect is formation of complex ions similar to those noted for the cadmium complexes. The tetrahedral symmetry of the $AlCl_4^-$ ion is regular in $AlCl_3$–NaCl melts and dis-torted in $AlCl_3$–KCl melts. In $ZnCl_2$–KCl melts, the concentration of discrete $ZnCl_4^{-2}$ ions increases with increasing amounts of KCl.[56c]

## SUMMARY

It is apparent that not infrequently the ionic and molecular species of the aqueous solution and the solid state persist in the in-organic molten state. The shifts of vibrational frequencies, splitting of degenerate fundamentals, and the appearance of "forbidden" fundamentals as noted in ionic melts appear to offer a direct approach to the evaluation of the environmental influences of the inorganic molten state. Chemical coordination-type interactions in mixtures of ionic and covalent salts have been confirmed* in the molten state. Quantitative contributions to the concept of partial covalency in the molten salts can be expected with the application of more sophisticated methods of vibrational analysis. High-temperature spectroscopy will assume increasing importance with the advance of inorganic chemistry in the molten state.

## REFERENCES

1. B. R. Sundheim (ed.), *Fused Salts*, McGraw-Hill Book Company, New York, 1964.
2. M. Blander (ed.), *Fused Salt Chemistry*, Interscience Publishers, Inc., New York, 1964.
3. G. J. Janz and R. D. Reeves, "Molten Electrolytes—Transport Processes," in: C. Tobias and P. Delahay (eds.), *Advances in Electrochemistry and Electrochemical Engineering, Vol. V*, Interscience Publishers, Inc., New York, 1965.
4. R. A. Bailey and G. J. Janz, "Experimental Techniques in the Study of Fused Salts," Chapter 7 in: J. J. Lagowski (ed.), *Chemistry of Non-Aqueous Solvents, Vol. 1*, Academic Press, Inc., New York, 1965.
5. R. Littlewood, *J. Electrochem. Soc.* **109**: 525 (1962).
6. M. D. Ingram and G. J. Janz, *Electrochim. Acta* **10**: 783 (1965).
7. J. D. MacKenzie, in: J. O'M. Bockris, J. L. White, and J. D. MacKenzie (eds.), *Physicochemical Measurements at High Temperatures*, Butterworths, London, 1959, Appendix XXXII.
8. J. L. White, in: J. O'M. Bockris, J. L. White, and J. D. MacKenzie (eds.), *Physicochemical Measurements at High Temperatures*, Butterworths, London, 1959, Appendix IV.
9. D. T. Livey and P. Murray, in: J. O'M. Bockris, J. L. White, and J. D. MacKenzie (eds.), *Physicochemical Measurements at High Temperatures*, Butterworths, London, 1959, Chapter 4.
10. I. E. Campbell, *High Temperature Technology*, John Wiley & Sons, Inc., New York, 1956.
11. H. J. Gardner, C. T. Brown, and G. J. Janz, *J. Phys. Chem.* **60**: 1459 (1956).
12. H. A. Laitinen, W. S. Ferguson, and R. A. Osteryoung, *J. Electrochem. Soc.* **104**: 516 (1957).
13. D. L. Hill, J. Perano, and R. A. Osteryoung, *J. Electrochem. Soc.* **107**: 698 (1960).
14. C. R. Boston and G. P. Smith, *J. Phys. Chem.* **62**: 409 (1958).
15. F. Maslan (ed.), *Nucl. Eng. Dept. Progr. Rept. BNL* 506 (1958).
16. Ya. E. Vil'nyanskii and E. I. Savinkova, *J. Appl. Chem. USSR (English Transl.)* **26**: 735 (1953).
17a. D. L. Maricle and D. N. Hume, *J. Electrochem. Soc.* **107**: 354 (1960).
17b. J. H. Freeman and M. L. Smith, *J. Inorg. Nucl. Chem.* **7**: 234 (1958).
18. H. V. Wartenberg, *Z. Anorg. Allgem. Chem.* **273**: 257 (1953).

---

* See also D. M. Gruen and co-workers, *J. Phys. Chem.* **63**: 393 (1959); *J. Inorg. Nucl Chem.* **4**: 74 (1957); *Nature* **178**: 1181 (1956); and A. R. Ubbelohde, *Proc. Chem. Soc. (London)* 332 (1960) for electronic absorption spectra and complex ions in fused salts.

19. F. R. Duke and A. S. Doan, Jr., *Iowa State J. Sci.* **32**: 451 (1958).
20. J. D. Corbett, S. V. Winbush, and F. C. Albers, *J. Am. Chem. Soc.* **77**: 3964 (1955).
21. G. Jander and R. Brodersen, *Z. Anorg. Allgem. Chem.* **262**: 33 (1950).
22. G. J. Janz, C. Baddiel, and T. R. Kozlowski, *J. Chem. Phys.* **40**: 2055 (1964).
23. C. C. Addison and N. Logan, in: N. J. Emelews and A. G. Sharpe (eds.), *Advances in Inorganic Chemistry and Radiochemistry*, Vol. 6, Academic Press, New York, 1964.
24. G. J. Janz, Y. Mikawa, and O. W. James, *Appl. Spectry.* **15**: 47 (1961).
25. J. U. White, L. Alpert, and A. G. De Bell, *J. Opt. Soc. Am.* **45**: 154 (1955).
26. L. A. Woodward, G. Garton, and H. L. Roberts, *J. Chem. Soc.* 3723 (1956).
27. T. F. Young and R. P. Westerdahl, U.S. Air Force, Office of Aerospace Research, Contract No. AF33(616)–5697, ARL Report No. 135 (1961).
28. W. Bues, *Z. Anorg. Allgem. Chem.* **279**: 104 (1955). See also J. O'M. Bockris, J. L. White, J. D. MacKenzie (eds.), *Physicochemical Measurements at High Temperatures*, Academic Press, Inc., New York, 1959.
29. N. S. Ham and A. Walsh, *Spectrochim. Acta* **12**: 88 (1958).
30. F. A. Miller and G. L. Carlson, *Spectrochim. Acta* **16**: 6 (1960).
31. F. T. King and E. R. Lippincott, *J. Opt. Soc. Am.* **46**: 661 (1956).
32. H. Stammreich, *Pure Appl. Chem.* **4**: 97 (1962).
33a. R. C. C. Leite and S. P. S. Porto, *J. Opt. Soc. Am.* **54**: 981 (1964).
33b. J. A. Koningstein and R. G. Smith, *J. Opt. Soc. Am.* **54**: 1061 (1964).
34a. L. J. Bellamy, *The Infrared Spectra of Polyatomic Molecules*, John Wiley & Sons, Inc., New York, 1954.
34b. G. M. Barrow, *Introduction to Molecular Spectroscopy*, McGraw-Hill Book Company, New York, 1954.
34c. R. P. Bauman, *Absorption Spectroscopy*, John Wiley & Sons, Inc., New York, 1962.
34d. G. Herzberg, *Molecular Spectra and Molecular Structure, II. Infrared and Raman Spectra of Polyatomic Molecules*, D. Van Nostrand Company, Inc., New York, 1945.
34e. G. J. Janz and Y. Mikawa, *J. Mol. Spectry.* **5**: 92 (1960).
34f. E. C. Curtis, *J. Mol. Spectry.* **17**: 108 (1965).
35. E. B. Wilson, Jr., J. C. Decius, and P. C. Cross, *Molecular Vibrations*, McGraw-Hill Book Company, New York, 1952.
36a. B. I. Stepanov and A. M. Prima, *Opt. i Spektroskopiya* **IV**: 734 (1958); **V**: 15 (1958).
36b. J. Zarzycki and F. Naudin, *J. Chim. Phys.* **58**: 830 (1961).
37. S. C. Wait, T. R. Kozlowski, and G. J. Janz, *J. Chem. Phys.* **39**: 1809 (1963).
38. G. W. Chantry and R. A. Plane, *J. Chem. Phys.* **32**: 319 (1960); **33**: 736 (1960); **35**: 1027 (1961).
39a. L. Coutoure, *Compt. Rend.* **220**: 646 (1945).
39b. M. Blander (ed.), *Fused Salt Chemistry*, Interscience Publishers, Inc., New York, 1964.
40a. R. A. Schroeder, C. E. Weir, and E. R. Lippincott, *J. Chem. Phys.* **36**: 2803 (1962).
40b. R. M. Hexter, *Spectrochim. Acta* **10**: 291 (1958).
40c. K. Buijs and C. J. H. Schutte, *Spectrochim. Acta* **18**: 307 (1962).
40d. F. Vrantny, *Appl. Spectry.* **13**: 59 (1959).
40e. B. M. Gatehouse, S. E. Livingstone, and R. S. Nyholm, *J. Chem. Soc.* 4222 (1957).
40f. J. R. Ferraro, *J. Mol. Spectry.* **4**: 99 (1960).
40g. W. E. Keller and R. S. Halford, *J. Chem. Phys.* **17**: 26 (1949).
40h. L. Coutoure, *Ann. Phys.* **2**: 94 (1947).
40i. O. Theimer, *Monatsh. Chem.* **81**: 424 (1950).
40j. L. P. Mathieu and M. Lounsbury, *Discussions Faraday Soc.* **9**: 196 (1950).

40k.  Y. Doucet and J. Vallier, *Compt. Rend.* **255**: 2935 (1962).
41a.  W. Bues, *Z. Physik. Chem. (Frankfurt)* **10**: 1 (1957).
41b.  J. K. Wilmshurst and S. Senderoff, *J. Chem. Phys.* **35**: 1078 (1961).
41c.  J. Greenberg and L. J. Hallgren, *J. Chem. Phys.* **33**: 900 (1960).
41d.  G. J. Janz and D. W. James, *J. Chem. Phys.* **35**: 739 (1961).
41e.  G. J. Janz, T. R. Kozlowski, and S. C. Wait, *J. Chem. Phys.* **39**: 1809 (1963).
41 f.  R. K. Khanna, J. Lingscheid, and J. C. Decius, *Spectrochim. Acta* **20**: 1109 (1963).
41g.  G. E. Walrafen and D. E. Irish, *J. Chem. Phys.* **40**: 911 (1964).
41h.  S. C. Wait, Jr., and A. T. Ward, *J. Chem. Phys.* **44**: 448 (1966).
 42.  G. P. Smith, "Electronic Absorption Properties of Molten Salts," in: M. Blander (ed.), *Selected Topics in Molten Salt Chemistry*, Interscience Publishers, New York, in press.
 43.  G. J. Janz and D. W. James, *Electrochim. Acta* **7**: 427 (1962).
 44.  K. Furukawa, *Discussions Faraday Soc.* **32**: 53 (1961).
 45.  J. Zarzycki, *Discussions Faraday Soc.* **32**: 212 (1961).
 46.  G. E. Walrafen, private communication (1965).
 47.  J. P. Devlin, K. Williamson, and G. Austin, *J. Chem. Phys.* (1965), in press.
48a.  H. Lee and J. K. Wilmshurst, *Australian J. Chem.* **17**: 943 (1964).
48b.  R. E. Hester and R. A. Plane, *Inorg. Chem.* **3**: 769 (1964).
49a.  J. K. Wilmshurst, *J. Chem. Phys.* **35**: 1800 (1961).
49b.  R. Buchanan, H. H. Caspers, and H. R. Marlin, *J. Chem. Phys.* **40**: 1125 (1964).
 50.  J. K. Wilmshurst, *J. Chem. Phys.* **36**: 2415 (1962).
51a.  G. E. Walrafen, *J. Chem. Phys.* **43**: 479 (1965).
51b.  G. E. Walrafen, D. E. Irish, and T. F. Young, *J. Chem. Phys.* **37**: 662 (1962).
 52.  S. T. Chen, Y. T. Yao, and T. Y. Wu, *Phys. Rev.* **51**: 235 (1937).
 53.  C. Rocchiccioli, *Compt. Rend.* **242**: 2922 (1956).
 54.  W. Bues, *Z. Anorg. Allgem. Chem.* **279**: 104 (1955).
55a.  W. Klemperer and L. Lindeman, *J. Chem. Phys.* **25**: 397 (1956).
55b.  A. Buchler, W. Klemperer, and A. G. Emslie, *J. Chem. Phys.* **36**: 2499 (1962).
56a.  E. J. Salstrom and L. Harris, *J. Chem. Phys.* **3**: 241 (1935).
56b.  D. E. Irish and T. F. Young, *J. Chem. Phys.* **43**: 1765 (1965).
56c.  J. R. Moyer, J. C. Evans, and G. Y. S. Lo, *J. Electrochim. Soc.* (1965), in press.
56d.  D. E. Irish, B. McCarroll, and T. F. Young, *J. Chem. Phys.* **39**: 3436 (1963).
 57.  K. Balasubrahmanyam, *J. Chem. Phys.* (1965), in press.
 58.  G. J. Janz and J. D. E. McIntyre, *Ann. N.Y. Acad. Sci.* **79**: 790 (1960); *J. Electrochem. Soc.* **109**: 842 (1962).
 59.  G. J. Janz and D. W. James, *J. Chem. Phys.* **38**: 902 (1963).
60a.  H. Braune and G. Engelbrecht, *Z. Physik. Chem. (Leipzig)* **B19**: 303 (1932).
60b.  K. V. K. Rao, *Proc. Indian Acad. Sci.* **A14**: 521 (1941).
60c.  E. J. Salstrom and L. Harris, *J. Chem. Phys.* **3**: 241 (1935).
 61.  G. Allen and E. Warhurst, *Trans. Faraday Soc.* **54**: 1786 (1958).
 62.  J. A. Rolfe, D. Sheppard, and L. A. Woodward, *Trans. Faraday Soc.* **50**: 1275 (1954).
63a.  L. A. Woodward, G. Garton, and H. L. Roberts, *J. Chem. Soc.* 3273 (1956).
63b.  L. A. Woodward, N. N. Greenwood, J. R. Hall, and I. J. Worrall, *J. Chem. Soc.* 1505 (1958).
63c.  J. Nixon and R. A. Plane, *J. Am. Chem. Soc.* **84**: 4445 (1962).
63d.  C. Baddiel and G. J. Janz, *Trans. Faraday Soc.* **60**: 2009 (1964).
 64.  W. Bues, *Z. Physik. Chem. (Frankfurt)* **10**: 1 (1957).
 65.  G. J. Janz and D. W. James, *J. Chem. Phys.* **35**: 739 (1961).
 66.  O. Borgen, K. Grjotheim, and S. Urnes, *Glastech. Ber.* **33**: 52 (1960).
67a.  G. J. Janz and D. W. James, *J. Chem. Phys.* **38**: 905 (1955).
67b.  W. Bues, *Z. Anorg. Allgem. Chem.* **279**: 104 (1955).

67c. M. A. Bredig and E. R. Van Artsdalen, *J. Chem. Phys.* **24**: 478 (1956).

67d. H. Gerding, H. G. Haring, and P. A. Renes, *Rec. Trav. Chim.* **72**: 78 (1953).

67e. H. Gerding and E. Smit, *Z. Physik. Chem. (Frankfurt)* **B50**: 171 (1941).

67f. G. J. Janz and T. R. Kozlowski, *J. Chem. Phys.* **39**: 843 (1963).

67g. K. Balasubrahmanyam and L. Nanis, *J. Chem. Phys.* **42**: 676 (1965).

68a. L. G. Sillen, *Acta Chem. Scand.* **3**: 549 (1949).

68b. R. C. Aggarwal, *Z. Physik. Chem. (Leipzig)* **207**: 1 (1957); *Z. Anorg. Allgem. Chem.* **291**: 134 (1957).

68c. T. R. Griffiths and M. C. R. Symons, *Trans. Faraday Soc.* **56**: 1752 (1960).

69. M. Rolla, P. Franzosini, and R. Riccardi, *Discussions Faraday Soc.* **32**: 84 (1961).

70. J. D. Corbett, *Inorg. Chem.* **1**: 700 (1962).

71a. M. A. Bredig, *Electrochim. Acta* **5**: 299 (1961).

71b. N. K. Boardman, F. H. Doran, and E. Heymann, *J. Phys. Chem.* **53**: 375 (1949).

71c. J. L. Barton and H. Bloom, *Trans. Faraday Soc.* **55**: 1792 (1959).

71d. M. Tanaka, K. Balasubramanyam, and J. O'M. Bockris, *Electrochim. Acta* **8**: 621 (1964).

72. M. A. Bredig, *J. Chem. Phys.* **37**: 451 (1962).

73. S. C. Wait, A. T. Ward, and G. J. Janz, *J. Chem. Phys.* **45**: 133 (1966).

*Chapter 6*

# Observed Resonance Raman Spectra

Josef Behringer

*Physikalisches Institut der Phil.–Theol. Hochschule
Eichstätt, Bayern, Germany*

## INTRODUCTION

Raman (R) spectra observed when the exciting frequency approaches or enters the region of electronic (vibronic) absorption of the molecule are called *resonance Raman* (RR) *spectra.* This condition is contrary to one of the conditions necessary for application of Placzek's polarizability theory (*cf.*, the work of Placzek,[11] p. 270) so that the resonance Raman effect (RRE) requires special theoretical treatment. In considering the vibrational RRE (to which we shall restrict ourselves in this article), it is convenient to distinguish between the *preresonance* case and the *rigorous resonance* case. In the former, the exciting frequency still lies outside the *observable* vibrational structure of the electronic absorption band responsible for R scattering (either on the long- or short-wave side); in the latter, it falls into the interior of this structure, either coinciding or not coinciding with a vibrational band. The fact that the vibrational structure of the absorption band may be latent (e.g., because of line broadening due to intermolecular interactions) does not seriously impair the above distinction, since in this case the characteristic features of the rigorous RRE (differing much from those of the preresonance effect) disappear. The limit between preresonance and ordinary Raman effect (RE) is completely arbitrary; every RE may be conceived as a rudimentary RRE.

By far, most of the investigations made in the past were devoted to the preresonance RE in liquid solutions and crystal powders. (For historical notes, see Behringer and Brandmüller,[17] p. 664 *ff* ; and Harrand and Martin.[35]) It is only recently that experimental data have been obtained on the rigorous RRE in liquids (see the section entitled "Rigorous Resonance Raman Effect," pp. 208–219).

The *general* theory of the effect is rather well developed, but its application to the interpretation of individual experimental results meets with extreme difficulties. The reasons are the complexity of the theoretical equations, which makes incisive simplifications unavoidable, and the insufficiency of existing data on excited states of polyatomic molecules.

In the following, we shall first consider some special features of the experimental technique and then give a detailed account of results obtained for the preresonance and the rigorous RRE, including only those details of theory which are indispensable for understanding.

## SPECIAL FEATURES OF EXPERIMENTAL TECHNIQUE

### General Remarks on the Observation of RR Spectra

Although resonance excitation raises the scattering power of many substances by several orders of magnitude, the RRE cannot always be observed with ease, because the scattered radiation may be abnormally attenuated by the simultaneously intensified absorption and possibly concealed by superimposed fluorescence emission. Moreover, photolysis may occur and in the course of time not only destroy the sample but also eventually produce unknown substances with different solubility and absorbance. Conditions for observability differ greatly among individual compounds. Even with a conventional arrangement, some substances yield RR spectra of such brightness as to be visible to the naked eye; others, however, require optimal performance of the instrumentation. In principle, the RRE does not require special instrumental equipment. For quantitative intensity measurements, any ordinary R spectrograph may be used, provided that it permits sufficiently accurate determinations of relative integrated band intensities. For detailed information (also concerning necessary corrections to measured intensity values due to varying instrumental factors), the reader is referred to the literature (see Brandmüller and Moser,[1] p. 235 *ff*, Geppert,[2] Bazhulin and Sushchinsky,[16] and Harrand and Martin[35]).

As already mentioned, the particular experimental difficulties in observing RR spectra are caused by absorption, fluorescence, and photolysis. These difficulties may always be overcome by choosing sufficiently low excitation frequencies. However, since this is synonymous with abandoning the condition for producing RR spectra, it is not relevant to the present discussion. As long as absorption and fluorescence may be traced back to impurities in the sample, they may, of course, be removed by the usual methods of sample preparation (see Brandmüller and Moser,[1] p. 219 *ff*, and Kohlrausch,[5] p. 57 *ff*). However, in this chapter we are primarily concerned with absorption and fluorescence derived from the sample substance itself and inseparably associated with the internal mechanism of the scattering process. These effects are still imperfectly understood and investigation of them is one of the most important aims in RR spectroscopy.

Before proceeding to the discussion of these questions intrinsic to the RRE, we shall give a short account of the equations for practical absorption corrections to be made to observed intensity values. Since the influence of absorption on the exciting and scattered radiation is negligible in the ordinary RE, the following section is to be understood as complementary to the literature on intensity correction factors in ordinary R spectroscopy.

### Influence of Absorption on Observed Band Intensities

**General.** In what follows we shall restrict ourselves to equations which have been proven successful for elimination of the influence of absorption on scanned band intensities of liquid solutions. Presumably, they would be equally applicable to gases. For extinction corrections on crystal-powder spectra, see Behringer and Brandmüller,[22] Melamed,[51] Moser and Varchmin,[55] and Vratny and Fischer.[88] The time-dependent variation of absorption resulting from photochemical transformations has been treated by Behringer,[19] p. 550 *ff*. Empirical methods for absorption correction are described by Rea[57] and Tsenter and Bobovich.[86, 87]

During passage through the scattering substances both the exciting and the scattered light undergo absorption which depends upon wavelengths and distances traversed. Let the scattering and simultaneously absorbing liquid be contained in a scattering vessel of volume $V$. A total radiation energy $I_0$ of uniform (practically monochromatic) frequency $v_0$ enters per $cm^2$-sec through the surface of the vessel, either by parallel or by diffuse incidence. (We only consider arrangements where $I_0$ is at least approximately constant over the whole surface.) Further, let $I_\varepsilon$ be the total scattered energy of definite and practically monochromatic frequency $v_\varepsilon$ (shifted or nonshifted relative to $v_0$) which emerges per sec from $V$ and is instrumentally detected. Then, certain types of scattering system geometries exist, some of which will be described below, which allow the influence of absorption on the exciting radiation to be exactly (Example 1 below) or approximately (Example 2) separated from its influence on the scattered radiation, so that the total influence of absorption may be expressed by the correction equation

$$I_\varepsilon = c_1 I_0 g(\chi_0) h(\chi_\varepsilon) \tag{1}$$

where the correction factors $g(\chi_0)$ and $h(\chi_\varepsilon)$ contain the absorption coefficients $\chi_0$ and $\chi_\varepsilon$ for the exciting and scattered frequencies $\nu_0$ and $\nu_\varepsilon$, respectively. The factor $c_1$ no longer contains $\chi_0$ and $\chi_\varepsilon$; it depends on the properties of the molecule and on other details of instrumentation.

When putting the equations given below to practical use, it is very important to note that the $\chi$ may be written (omitting indices) as follows:

$$\chi = -\frac{1}{I}\frac{dI}{dl} = \epsilon c \ln 10 \tag{2}$$

where $\epsilon$ is the decadic molar absorption coefficient and $c$ is the concentration of absorbing substance in moles/liter. If in a solution it is essentially only the light-scattering solute that absorbs both exciting and scattered radiation, it may be shown from the explicit forms of equation (1) that an "optimal concentration" $c_{opt}$ exists, yielding the most advantageous ratio of scattered to absorbed energy. The term $c_{opt}$ may be found easily by maximization of $I_\varepsilon = I_\varepsilon(c)$. We shall now consider two special examples of the idealized arrangements fulfilling equation (1).

**Example 1.** Let the primary radiation in a direction parallel to $x$ (Fig. 1) fall upon a parallelepiped scattering volume $V = x_0 y_0 z_0$ and let only the radiation scattered in direction $+z$ and leaving the front face $x_0 y_0$ at $z = z_0$ be collected by the light-receiving system. Then we have exactly

$$I_\varepsilon = c_1 I_0 c y_0 \frac{1 - e^{-\chi_0 x_0}}{\chi_0} \frac{1 - e^{-\chi_\varepsilon z_0}}{\chi_\varepsilon} \tag{3}$$

(cf., Vratny and Fischer,[13] Behringer,[19] p. 546, Fisher and Lippincott,[32] Fisher,[33] Kecki,[41] Lippincott et al.,[48] Michel,[53] and Vratny[89]).

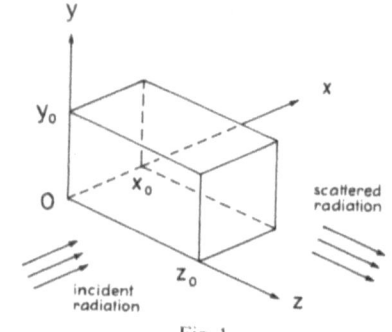

Fig. 1.

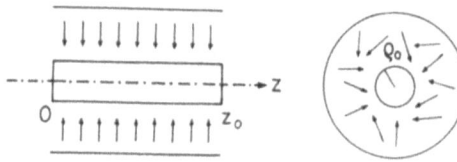

Fig. 2.

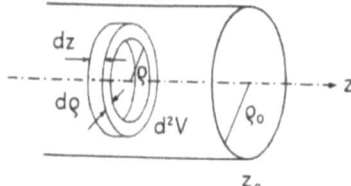

Fig. 3.

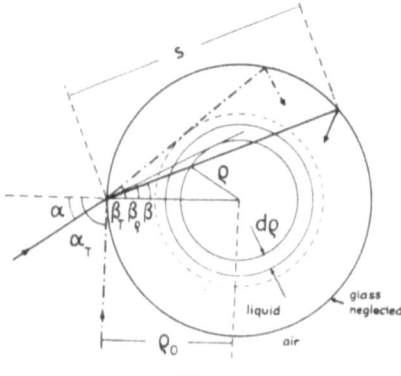

Fig. 4.

**Example 2.** A cylindrical R tube (length $z_0$, radius $\rho_0 \ll z_0$) is irradiated in planes perpendicular to the tube axis (rotation-symmetrical arrangement, idealized helical lamp with reflector and baffles); within the planes, incidence is diffuse (Fig. 2).

The exciting radiation density $I_0(\rho)$ within the ring-shaped volume element $d^2V = 2\rho\pi\,d\rho\,dz$ (Fig. 3) depends only on $\rho$ and may be exactly calculated (see below). The scattered radiation produced in $d^2V$ is propagated in different, and very complicated ways, and it partially reaches the tube window at $z = z_0$. The usual angular apertures of spectrographs with or without optical imaging systems permit the assumption that without serious error the radiation reaching the detecting system traverses the tube in a direction approximately parallel to the $z$-axis. This assumption may still be maintained if the observed intensity is collected to a certain extent by total reflection within the tube, which is very largely the case for small tube diameters[26] and which because of intensity gains may even be considered as advantageous for the observation of the RRE. Under these circumstances, the attenuating factor for the *scattered* radiation by integration along $z$ becomes approximately

$$h(\chi_\varepsilon) = \frac{1 - e^{-\chi_\varepsilon z_0}}{\chi_\varepsilon} \tag{4}$$

as in equation (3). Numerous experimental checks have shown that equation (4) is valid to an excellent approximation (*cf.*, Schrötter[80,81] and references mentioned under Example 1).

We now give a very general equation for the attenuation of the *incident* light by absorption. Neglecting the influence of the glass walls of the R tube (which for small glass thickness practically leads to no error), one gets after a somewhat laborious calculation (Fig. 4)

$$I_0(\rho) = I_0\frac{n_0}{2\pi}\int \frac{e^{-\chi_0\rho_0\cos\beta}\left(e^{\chi_0\sqrt{\rho^2 - \rho_0^2\sin^2\beta}} + e^{-\chi_0\sqrt{\rho^2 - \rho_0^2\sin^2\beta}}\right)(1 - R)}{(1 - Re^{-2\chi_0\rho_0\cos\beta})\sqrt{\rho^2 - \rho_0^2\sin^2\beta}}$$

$$\times \rho_0\cos\beta\,d\beta \tag{5}$$

where $n_0$ is the refractive index at $\nu_0$ and $R$ is the reflectivity at the liquid–air boundary dependent on $\alpha$, $n_0$, and polarization according to the Fresnel equations. With $\beta_T = \arcsin 1/n_0$, the limiting value of the total reflection angle and $\beta_\rho = \arcsin \rho/\rho_0$, the tangential angle for the ring radius $\rho$ (Fig. 4), the integration in equation (5) has to be extended

$$\text{for } \beta_\rho < \beta_T \quad \text{from} \quad -\beta_\rho \quad \text{to} \quad +\beta_\rho$$

and

$$\text{for } \beta_\rho > \beta_T \quad \text{from} \quad -\beta_T \quad \text{to} \quad +\beta_T$$

Equation (5) includes the absorption, refraction, and multiple reflection effect on the primary radiation under the special conditions described. In the case of vanishing absorption ($\chi_0 = 0$), it contains the results found by Rea[57] that (a) reflections are then of no importance and (b) there is a homogeneously irradiated cylindrical zone of radius $\rho_0/n_0$ in the interior of the tube. The scattered intensity excited in $d^2V$ is proportional to $I_0(\rho)\,d^2V$. If the integration is carried out from $\rho = 0$ to $\rho_0$, the scattered intensity excited in a tube cross section of thickness $dz$ is found to be proportional to $I_0\,dzg(\chi_0)$ where

$$g(\chi_0) = \rho_0 \int_{-\pi/2}^{+\pi/2} \frac{1 - R(\alpha)}{1 - R(\alpha)e^{-\chi_0 s(\alpha)}} \frac{1 - e^{-\chi_0 s(\alpha)}}{\chi_0} \cos\alpha\,d\alpha \qquad (6)$$

where $\sin\alpha = n\sin\beta$ and $s(\alpha) = 2\rho_0\cos\beta$ (cf., Behringer[10]).

## PRERESONANCE RAMAN EFFECT

### Prefatory Remarks on Observed Spectra

**Intensity Scale Used.** Unless otherwise specified, we shall use throughout this section relative integrated band intensity values referred to a scale in which the 313 cm$^{-1}$ band of carbon tetrachloride is adopted as reference standard and is given 100 intensity units (IU). This intensity scale was introduced by Shorygin.[62;63;64, p.425] The above definition presupposes identical operating conditions for the sample and the reference substance, i.e., use of the same excitation frequency, particle number density, illumination, and detection, and due application of the necessary intensity corrections for instrumental and absorption factors. The simplest way to guarantee identical operating conditions with sufficient accuracy is to use carbon tetrachloride or another appropriate reference substance (previously related for intensity normalization to carbon tetrachloride) as internal standard in gaseous, liquid, or solid mixtures, finally reducing measured band intensities of specimen and standard to the same molar concentrations. However, with this method special note should be taken of solvent effects in liquid or solid mixtures which occasionally destroy the linearity of observed intensity dependence on relative concentrations to a considerable extent. The intensity values given

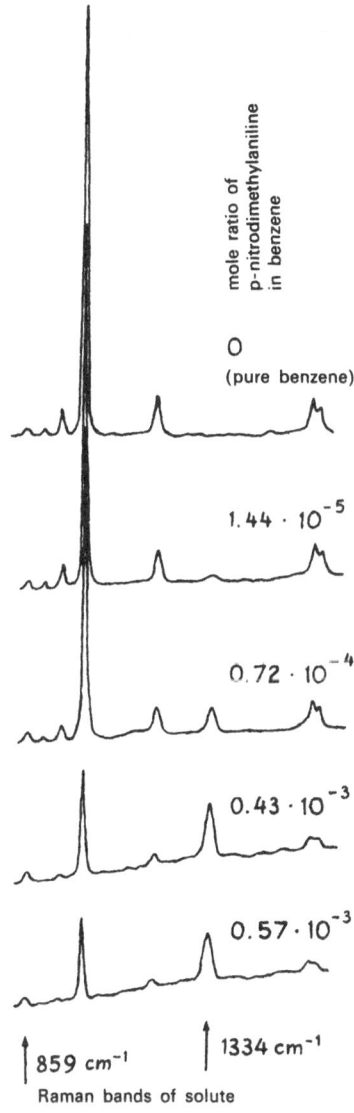

Fig. 5. RRE of p-nitrodimethylaniline in benzene solution after Matz, cited in the work of Brandmüller and Moser,[27] p. 582.

in the following sections may be taken on the average to be subject to an uncertainty of not more than 20%. We shall quote errors when they were mentioned in the original papers. Detailed description for photoelectric recording and evaluation methods of R band intensities including absorption corrections are given in Brandmüller and Moser,[1] p. 235 ff, Behringer,[19] p. 551 ff, and Maier.[49,50]

**Examples of Recorded Resonance Raman Spectra.** In principle, the RRE can be observed in the gaseous, liquid, and solid states. RR spectra of rarefied gases have been known for a long time as "resonance spectra of fluorescence" and will not be considered here. Many photoelectric measurements have been made on liquid solutions under resonance excitation. For a given exciting frequency, there are many substances showing the RRE with intensity sufficient for observation only when concentrations are used close to the optimal concentration (cf., section entitled "General," p. 171). This may be seen from Figs. 5 and 6. For solutions of p-nitrodimethylaniline in benzene or methanol, the optimal concentrations (given as mole ratio of solution, i.e., number of moles of dissolved sample per number of moles of solvent) are $0.5 \times 10^{-3}$ and $4.1 \times 10^{-6}$, respectively, when Hg e is used as the exciting line. Up to the present, RR spectra of crystal powders have been mostly recorded by the photographic method, which does not yield intensity values of comparable accuracy (see Behringer and Brandmüller,[22] Harrand,[34,36] and Théry et al.;[84] photoelectric recording in Bobovich and Eidus[25]). However, the general features of the RRE of solids are indubitably the same as those for liquids.

## Characteristic Features

**High-Intensity Values.** It may be recognized from Figs. 5 and 6 that extremely small concentrations of the sample in liquid solution are detectable by the RRE. This is just another consequence of the frequently enormous rise of molecular scattering power which may lead to intensity values greater by several orders of magnitude than those observed under normal exciting conditions. In the example mentioned above, the resulting intensity value computed for the symmetric $NO_2$-valence vibration R band $\tilde{v}(\pi) \approx 1310 \, \mathrm{cm}^{-1}$ is 270,000 IU or 1,500,000 IU ($\pm 10\%$) for p-nitrodimethylaniline in benzene or methanol solutions, respectively,[80] p. 863. Nearly all examples to be given below will demonstrate this enhancement of band intensities typical of the RRE, so that for purposes of exemplification we may confine ourselves here to the most striking cases.

Figure 7 shows the RR spectrum of $\beta$-carotene excited with Hg c.[21] The two bands at 1521 and 1155 cm$^{-1}$ have intensity values of $5 \times 10^6$ and $10^7$ IU ($\pm 50\%$), respectively. Figure 8 presents the

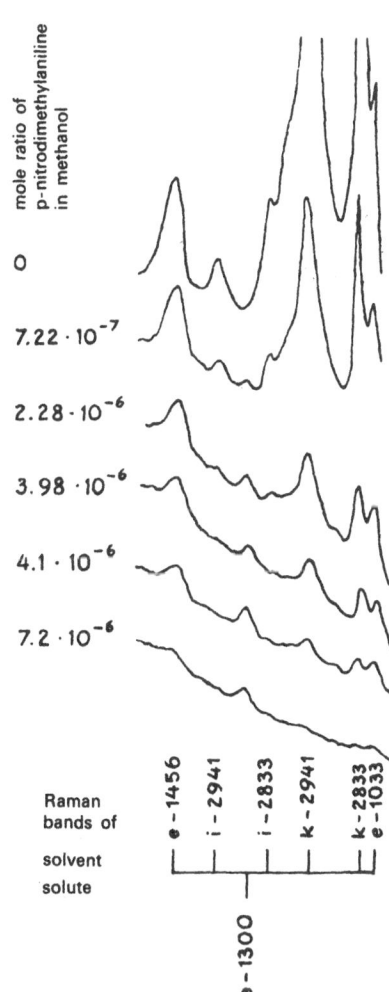

Fig. 6. RRE of p-nitrodimethylaniline in methanol solution.[19] Steinheil GH Universal Spectrograph (three prisms; collimator lens, $f = 650$ mm; camera lens, $f = 640$ mm) with photoelectric recording system. Entrance and exit slit widths, 100 $\mu$ (spectral slit width 5 cm$^{-1}$ at Hg e). Recording speed, 1 mm/min; time constant, 2 sec; dynode supply voltage, 1200 V on RCA 1P21 multiplier. Toronto-type mercury low-pressure lamp; supply, 87 V and 15.3 A DC. Sample temperature, 16°C. No filters.

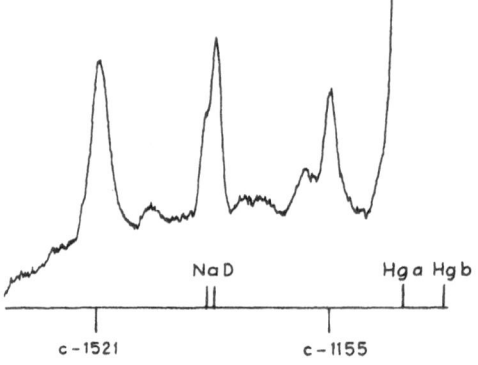

Fig. 7. RR spectrum of β-carotene in carbon tetrachloride solution.[18,21] Mole ratio 1 : 8190. Recording conditions as stated for Fig. 6 with one exception : recording speed, 2 mm/min.

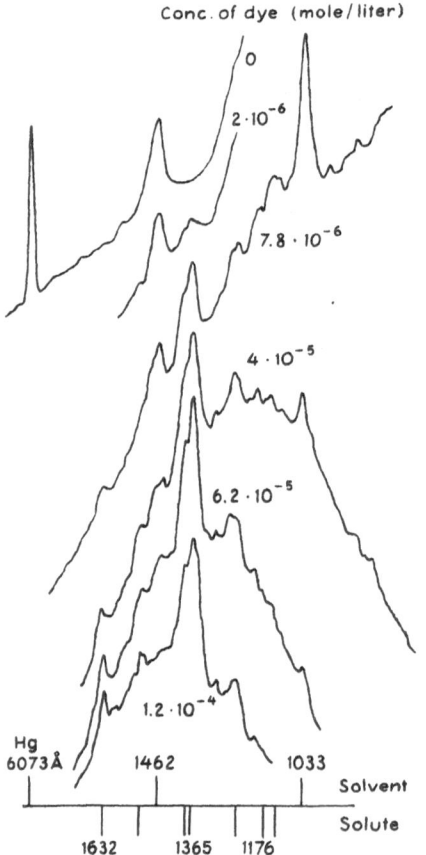

Fig. 8. RR spectrum of pseudo-isocyanine in methanol solution after Maier and Dörr.[49,50] Instrumentation similar to Fig. 6. Spectral slit width, $10\,cm^{-1}$; dynode supply voltage, 900 V; recording speed, 1 cm/sec; time constant, 2 sec. Neodymium filter solution.

RR spectrum of pseudo-isocyanine excited with Hg c after the method of Maier and Dörr.[49] The intensity of the band at $1365 + 1395\,\mathrm{cm}^{-1}$ amounts to $46 \times 10^6\,\mathrm{IU}\,(\pm 9\%)$. A still higher intensity value, $115 \times 10^6\,\mathrm{IU}\,(\pm 9\%)$, was found by the same authors for the 5,6-benzo-6'-chloro-pseudo-isocyanine band at $1358\,\mathrm{cm}^{-1}$ under similar operating conditions. The highest RR intensities so far were measured by Shorygin and Ivanova,[77] who discovered that the diphenyl-dodecahexene bands at 1142 and $1552\,\mathrm{cm}^{-1}$ excited by Hg e in acetone solution have the values $35 \times 10^7$ and $53 \times 10^7$ IU, respectively.

These intensity values are rather impressive; however, the reader is warned against the misconception that the substances mentioned should necessarily show R spectra of beaming brightness, the surprising fact not being the apparent splendor of R lines but the low values of the concentrations at which they can still be observed.

### Deviations from the $\nu_\varepsilon^4$-Law

*Theoretical Aspects.* It does not make much sense to use the concept of characteristic intensities (see Brandmüller and Moser,[1] p. 322, and Behringer and Brandmüller,[17] p. 659) in the RRE, since the R band intensities in the resonance region are strongly dependent on the position of the exciting frequency relative to the absorption band maxima of the molecule. It is the proximity of the proper electronic or vibronic absorption frequencies of the molecule which is responsible for the enhancement of intensity values as well as of intensity rates of change. The variation of R band intensities with choice of exciting frequency is theoretically more important than their formally impressive magnitudes. This intensity variation may lead to conclusions on molecular structure inaccessible to the normal RE.

By quantum theory, the total intensity (solid angle $4\pi$) of frequency $\nu_\varepsilon = \nu_0 \pm \nu = \nu_0 + \nu_{mn}$ (nuclear frequency $\nu = |\nu_{mn}| > 0$, $\nu_{mn} = -\nu_{nm} = \nu_m - \nu_n$) scattered on the average by one freely orientable molecule passing from state $m$ to state $n$ is given by [20]

$$I_{mn} = \frac{2^7 \pi^5}{3^2 c^4} I_0 \nu_\varepsilon^4 \sum_{\rho,\sigma} |\alpha_{\rho\sigma,mn}|^2 \tag{7}$$

where $\alpha_{\rho\sigma,mn}$ is the $\rho\sigma$ component ($\rho, \sigma = X, Y, Z$ space-fixed Cartesian coordinate system) of the "scattering tensor" (dyadic)

$$\alpha_{mn} = \frac{1}{h} \sum_r \left( \frac{\mu_{rn}\mu_{mr}}{\nu_{rm} - \nu_0 + i\delta_r} + \frac{\mu_{mr}\mu_{rn}}{\nu_{rn} + \nu_0 + i\delta_r} \right) \tag{8}$$

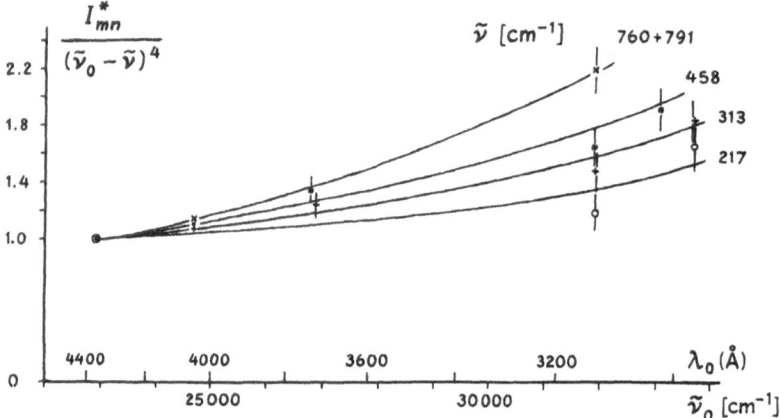

Fig. 9.  Deviations from $v_\varepsilon^4$-law for carbon tetrachloride R bands after Hofmann and Moser.[37]  Ordinate normalized to 1 for Hg e.  Exciting lines: Hg 4358, 4047, 3663, 3655, 3132, 3022, and 2967 Å.  Corrections made for refractive index after L. A. Woodward and J. M. B. George [*Nature* **167**: 193 (1951)] and those for absorption by means of equation (4).  Vertical short lines represent experimental errors.

where $c$ is the velocity of light; $h$ is Planck's constant; $I_0$ is the incident plane-polarized intensity of frequency $v_0$; $r$ is the intermediate state; $\mu_{rn}$ and $\mu_{mr}$ are the transition moments of the electronic dipole moment operator $\mu$; and $\delta_r$ is the damping constant of $r$.*  If the $v_0$-dependence of $\alpha_{mn}$ described by equation (8) is neglected, $I_{mn}$ depends on $v_0$ only through $v_\varepsilon^4$, which explains the expression "$v_\varepsilon^4$-law." The theoretical explanation of the RRE is basically connected with the "resonance denominators" in equation (8).  It may easily be understood that by approximation of $v_0$ toward a certain vibronic transition frequency $v_{rm}$ the corresponding term in equation (8) becomes dominant provided that the transition moments and damping constants fulfill certain conditions.  As a consequence, a more or less strong dependence of $\alpha_{mn}$ and, therefore, also of $I_{mn}/v_\varepsilon^4$ on $v_0$ results.  This deviation from the $v_\varepsilon^4$-law represents the theoretical basis for quantitative investigations of the RRE.  It may qualitatively be said that the more these deviations become noticeable the more a RRE takes place.

*Experimental Results.*  As early as in the 1930's Placzek discussed the $v_0$-dependence of scattering amplitudes by examination of the

---

* The double sum in equation (7) may also be written as $\alpha_{mn}:\alpha_{mn}^\dagger$, where : represents the scalar product of two tensors (or double-scalar product of two dyadics) defined by $\mathbf{ab}:\mathbf{cd} \equiv (\mathbf{a} \cdot \mathbf{d})(\mathbf{b} \cdot \mathbf{c})$ and where the symbol "$\dagger$" is the Hermitian adjoint (or conjugate complex transpose).

$v_\varepsilon^4$-law (see Placzek,[11] p. 368, and Behringer and Brandmüller,[17] p. 665). The measurements on carbon tetrachloride quoted by Placzek were repeated under improved experimental conditions (photoelectric recording) by Hofmann and Moser.[37] These authors determined the intensity ratio of the $CCl_4$ Raman bands to the exciting Hg lines lying in the region from 4358 to 2967 Å (intensity measured at the situation of the R tube by means of a reflecting small MgO-coated plate). The result is given in Fig. 9. There is for all lines a deviation from the $v_\varepsilon^4$-law which increases with $v_0$ and is notably different for the single R lines of carbon tetrachloride. Later the same authors[38] published the results of analogous measurements (the MgO plate was replaced by a MgO-powder suspension in water which filled the R tube) which are reproduced in Table I. The data for $CCl_4$ qualitatively correspond to those reported in Fig. 9, although the rate of increase (column 5) seems to be smaller for all lines (for $217\,cm^{-1}$ even $<1$). (For $CCl_4$ $459\,cm^{-1}$, cf., Tsenter and Bobovich.[87]) Many substances show

## TABLE I

### Deviation from $v_\varepsilon^4$-Law by Comparison of Excitation with Hg Lines 4358 and 2537 Å After Hofmann and Moser[38]

| Compound | Vibrational wave number $\tilde{v}$ (cm$^{-1}$) | Symmetry representation | Intensity ratio* $\dfrac{I_{R2537}}{I_{R4358}}$ | Deviation factor* $\dfrac{I_{R2537}}{I_{R4358}} \cdot \dfrac{(\tilde{v}_1 - \tilde{v})^4}{(\tilde{v}_2 - \tilde{v})^4}$ |
|---|---|---|---|---|
| $CCl_4$ | 459 | $A_1$ | 21 | 2.3 |
| (pure) | 217 | E | 8 | 0.9 |
| | 314 | $F_2$ | 12 | 1.4 |
| | 760/792 | $F_2$ | 30 | 3.2 |
| $Na_2SO_4$ | 981 | $A_1$ | 18 | 1.9 |
| (aq. soln.) | 451 | E | 23 | 2.5 |
| | 1104 | $F_2$ | 24 | 2.6 |
| $K_2CO_3$ | 1063 | $A_1'$ | 17 | 1.8 |
| (aq. soln.) | 1415 | $E'$ | 48 | 5.1 |
| $NaNO_3$ | 1050 | $A_1'$ | 160 | 17 |
| (aq. soln.) | 1390 | $E'$ | 410 | 42 |
| $KNO_2$ | 1331 | $A_1$ | 1220 | 124 |
| (aq. soln.) | 813 | $A_1$ | $850 \pm 400$ | $90 \pm 40$ |

* $I_{R4358}$ and $I_{R2537}$ are the Raman line intensities excited by Hg 4358 and 2537 Å, respectively. $\tilde{v}_1$ and $\tilde{v}_2$ are the wave numbers of Hg 4358 and Hg 2537 Å, respectively. $\tilde{v}$ is the wave number of the Raman line.

## TABLE II
### Deviation from the $\nu_\varepsilon^4$-Law After Khalilov[43]

| Compound | Formula | $I_{rel} = \dfrac{I_{\nu,\text{Sample}}}{I_{\nu,\text{Reference}}}$ | Exciting line | | | |
|---|---|---|---|---|---|---|
| | | | Hg c | Hg e | Hg k | Hg q |
| Ethyl ether of cinnamic acid | benzene–CH=CH–C(=O)–OC$_2$H$_5$ | $I_{1595}/I_{1442}$ | — | 1.17 | 1.42 | 2.66 |
| | | $I_{1631}/I_{1442}$ | — | 2.22 | 3.7 | — |
| Styrene | benzene–CH=CH$_2$ | $I_{1600}/I_{1442}$ | 0.59 | 0.72 | 0.93 | 1.31 |
| | | $I_{1630}/I_{1442}$ | 0.75 | 1.02 | 1.42 | — |
| Benzonitrile | benzene–C≡N | $I_{1597}/I_{1442}$ | 0.67 | 0.77 | 0.93 | 1.29 |
| | | $I_{2224}/I_{1442}$ | 1.38 | 1.81 | 2.23 | 2.58 |
| Nitrobenzene | benzene–NO$_2$ | $I_{1348}/I_{1442}$ | 0.95 | 1.71 | — | — |
| Crotonaldehyde | CH$_3$–CH=CH–CHO | $I_{1640}/I_{1442}$ | 0.43 | 0.97 | 1.98 | — |
| | | $I_{1690}/I_{1442}$ | 1.45 | 1.87 | — | — |
| Cyclohexane | C$_6$H$_{12}$ | $I_{1442}/I_{459\ \text{CCl}_4}$ | 1.04 | 1.20 | 1.07 | — |

considerably greater deviations from the $v_{\varepsilon}^4$-law than those of carbon tetrachloride. This may be seen from the data given for $NaNO_3$ and $KNO_2$ in Table I. Table II is a reproduction of a series of measurements made by Khalilov[42,43] on the intensity ratio (sample line/cyclohexane line 1442 cm$^{-1}$) by excitation with four different Hg lines. According to the last row, there is an approximate equality of behavior for cyclohexane 1442 cm$^{-1}$ and carbon tetrachloride 459 cm$^{-1}$, so that the intensity increases of the R bands with diminution of the exciting wavelength are approximately additional to those of carbon tetrachloride. It turns out that the deviations from the $v_{\varepsilon}^4$-law for visible excitation are smallest for substances whose electronic absorption band maxima are situated far away in the ultraviolet. This is the case for carbon tetrachloride, methanol, acetone, and cyclohexane, for example. On the other hand, the closer the electronic absorption bands approach the visible exciting frequency, the bigger those deviations will be.

Recognition of this leads to an alternative way of investigating the $v_0$-dependence of the R intensities: Instead of using several exciting frequencies for one individual compound, the experimental program may also be based upon the investigation by means of one exciting frequency of a series of structurally similar compounds with advancing long-wave absorption. This procedure presupposes, however, the existence of characteristic vibrations substantially independent of chemical substitutions or structure changes of the total molecule, and it is therefore liable to criticism. Although an exact quantitative examination of the $v_0$-dependence of R intensity cannot be effected by this method, it is this way which demonstrates most impressively the typical phenomenon of R intensity increase by mutual approach of absorption and exciting frequencies. We will consider some examples.

The para-substituted nitrobenzene derivatives listed in Table III were first investigated by Shorygin[66-68] (cf., Behringer,[19] Brandmüller et al.,[28] and Schrötter[80]). According to column 6, their very intense first absorption bands move progressively toward the visible spectral region. The last of them are intensively yellow-colored compounds. Excitation was by Hg e, with two exceptions that were by Hg k. Columns 4 and 5 contain the wave numbers and intensities, respectively, for the R band due to the symmetric stretching of the $NO_2$ group (see Fig. 10). Data for the first absorption band are given in columns 6–8.

## TABLE III
### Resonance Raman Effect of para-Substituted Nitrobenzene Derivatives After Shorygin[66-68]

| Compound | Formula | Solvent | $\bar{\nu}_{NO_2}$ (cm$^{-1}$) | $I_{NO_2}$ (IU) | First absorption band | | |
|---|---|---|---|---|---|---|---|
| | | | | | Maximum $\bar{\nu}_{eg}$ (cm$^{-1}$) | Distance from excitation line $\bar{\nu}_{eg} - \bar{\nu}_0$ (cm$^{-1}$) | Molar decadic absorption coefficient $\epsilon_{\nu_0}$ |
| Nitropropane | $CH_3$–$CH_2$–$CH_2$–$NO_2$ | Benzene | 1380 | 40 | — | — | — |
| Nitrobenzene | $NO_2$ (phenyl) | Benzene | 1348 | 900 | ~38,000 | ~15,000 | — |
| p-Nitrotoluene | $NO_2$—⬡—$CH_3$ | Cyclohexane | 1346 | 1,100 | 38,000 | 15,000 | 0.04 |
| | | Benzene | 1346 | 1,200 | ~36,500 | ~13,500 | 0.06 |
| | | Methanol | 1348 | 1,900 | 36,400 | 13,500 | 0.07 |
| p-Nitrophenol | $NO_2$—⬡—$OH$ | Benzene | 1344 | 4,000 | 33,200 | 10,000 | 0.08 |
| p-Nitroaniline | $NO_2$—⬡—$NH_2$ | Benzene | 1335 | 20,000 | 29,000 | 6,000 | 5 |
| p-Nitrodimethylaniline | $NO_2$—⬡—$N(CH_3)_2$ | Benzene | 1319 | 170,000 | 26,000 | 3,000 | 80 |
| | | Methanol | 1307 | 1,300,000 | 25,700 | 2,700 | 5,000 |
| p-Nitrodiethylaniline | $NO_2$—⬡—$N(C_2H_5)_2$ | Cyclohexane | 1330 | 70,000 | 27,600 | 4,600 | 5 |
| | | Benzene | 1318 | 300,000 | 25,700 | 2,700 | 170 |
| | | Methanol | 1307 | 2,000,000 | 25,000 | 2,000 | 8,000 |
| | | Methanol | 1307 | ~4,000,000* | 25,000 | 300 | 20,000 |
| K-p-nitrophenolate | $NO_2$—⬡—$O^-$ $K^+$ | Water + methanol | 1285 | ~1,000,000* | 24,850 | 150 | 19,000 |

* Excitation by Hg k is indicated by an asterisk; unlabeled values refer to excitation by Hg e.

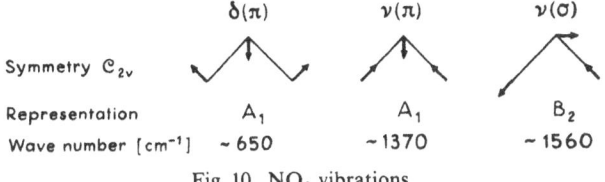

Fig. 10. $NO_2$ vibrations.

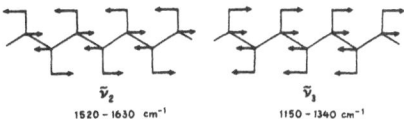

Fig. 11. Polyene-chain vibrations (schematic).

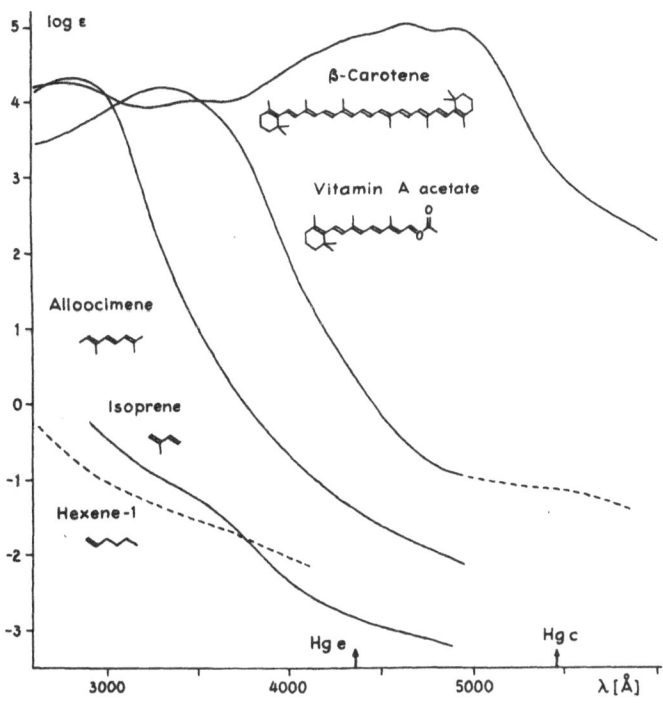

Fig. 12. Absorption spectra of polyenes in carbon tetrachloride solutions.[18]

## TABLE IV
### Resonance Raman Effect of Polyene Compounds [19]

| Compound | First absorption band maximum $\lambda_{eg}^{max}$ (Å) | Number of C=C bonds | $\tilde{v}_1$ (cm⁻¹) | $I_1$ (IU) | $\tilde{v}_2$ (cm⁻¹) | $I_2$ (IU) | $\tilde{v}_3$ (cm⁻¹) | $I_3$ |
|---|---|---|---|---|---|---|---|---|
| Hexene-1 | ≪2400 | 1 | 3000 | 70 | 1638 | 38 | 1300 | <120 |
| Isoprene | <2400 | 2 | 3008 | 49 | 1637 | 272 | 1290 | 97* |
| Alloocimene | 2800 | 3 | 3010 | 50 | 1629 | 5,560 | 1235 | 520 |
| Vitamin-A acetate | 3320 | 5 (+C=O) | ? | ? | 1580 | 45,000 | 1189 | 9,500 |
| β-Carotene | ~4600 | 11 | ? | ? | 1521 | 10,000,000* | 1155 | 5,000,000* |

* Error of the intensity value is ±20% except for those values marked by an asterisk where it is ±50%.

## TABLE V
### Resonance Raman Effect of Diethyl Ethers of Polyene-$\alpha,\omega$-Dicarbon Acids $C_2H_5OOC(CH=CH)_nCOOC_2H_5$ ($n = 1, \ldots, 5$) After Ivanova, Yanovskaya, and Shorygin [39]

| $n$ | $\tilde{v}_3$ (cm⁻¹) | $I_{\tilde{v}_3}$ (IU) | $\tilde{v}_2$ (cm⁻¹) | $I_{\tilde{v}_2}$ (IU) | $\tilde{v}_{CO}$ (cm⁻¹) | $I_{\tilde{v}_{CO}}$ (IU) | Absorption band first maximum $\tilde{v}_{eg}$ (cm⁻¹) | Absorption coefficient $\epsilon_{max}$ |
|---|---|---|---|---|---|---|---|---|
| 1 | 1210 | 50 | 1664 | 300 | 1733 | 330 | 47,400 | 16,600 |
| 2 | 1135 | 800 | 1644 | 5,000 | 1726 | 800 | 37,800 | 30,000 |
| 3 | 1139 | 7,000 | 1621 | 30,000 | 1721 | 1800 | 31,750 | 40,500 |
| 4 | 1139 | 44,000 | 1596 | 150,000 | 1717 | 6000 | 28,500 | 57,600 |
| 5 | — | — | 1570 | 1,100,000 | — | — | 26,050 | 66,600 |

The polyene compounds specified in Table IV (Behringer,[19] p. 562) in carbon tetrachloride solutions were excited by Hg e, with the exception of $\beta$-carotene for which Hg c was used (Fig. 7). Figure 11 is a schematic of the vibrational modes $\nu_2$ and $\nu_3$ of the polyene chain, the R band intensities of which according to Table IV (columns 5, 7, and 9) are successively considerably more enhanced than that of the CH-valence vibration $\nu_1$. Figure 12 presents the absorption spectra of these polyenes (the arrows indicate the positions of the exciting frequencies).

Similar compounds are the diethyl ethers of the polyene-$\alpha,\omega$-dicarbon acids listed in Table V. Here again, the measurements on acetone solutions by excitation with Hg e performed by Ivanova, Yanovskaya, and Shorygin[39] show a distinct inequality of intensity variation of the polyene-chain bands $\nu_2$ and $\nu_3$ compared with the C=O-valence vibration band $\nu_{CO}$. (For normal vibration calculations of the polyene chain, cf., Popov and Kogan.[56])

In summary, we can state that the theoretically expected augmentation of R band intensities, caused by relative approach of electronic absorption and excitation frequencies and surmounting the $\nu_\varepsilon^4$-proportionality, is indeed generally observed. In the following, we shall examine this effect more in detail.

### Effective Absorption Frequencies

**Semi-Classical Theory of Shorygin.** In order to attain quantitative conclusions from the above experimental findings and other known molecular data, the rather complicated equations of the general theory have to be considerably simplified. We first replace the indices $m$, $r$, and $n$ in equation (8) by the double indices $gv_1$, $ev$, and $gv_2$, respectively, for vibronic states ($g$ and $e$ are quantum numbers for pure electronic ground and excited states; $v_1$, $v$, and $v_2$ are quantum numbers for vibrational sublevels of initial, intermediate, and final vibronic states), thus taking into account only a ground-state RE. Omission of the "infrared term" in the electronic summation index $g$ (small for $\nu_0 \gg$ ground state vibrational frequencies $\nu = |\nu_{gg}^{v_1 v_2}|$) yields

$$\alpha_{gg}^{v_1 v_2} = \frac{1}{h} \sum_{e \neq g} \sum_v \left( \frac{\mu_{eg}^{v v_2} \mu_{ge}^{v_1 v}}{\nu_{eg}^{v v_1} - \nu_0 + i\delta_e^v} + \frac{\mu_{ge}^{v_1 v} \mu_{eg}^{v v_2}}{\nu_{eg}^{v v_2} + \nu_0 + i\delta_e^v} \right) \qquad (9)$$

We now assume $\nu_0$ is sufficiently close to the transition frequencies $\nu_{e_1 g}^{v v_1}$ of the first electronic state $e = e_1$ (which shall be nondegenerate) and all higher electronic states are sufficiently distant from $e_1$ so that the sum over $e$ may be reduced to the one term with $e = e_1$. Furthermore (for the present), we assume $\delta_e^v$ is negligibly small. If, moreover, $v_0 \gg |v_{gg}^{v_1 v_2}|$, we have approximately

$$v_{eg}^{vv_1} \approx v_{eg}^{vv_2} \tag{10}$$

so that for a real and *eo ipso* symmetric scattering tensor it follows (again with use of $e$ instead of $e_1$) that

$$\alpha_{gg}^{v_1 v_2} = \sum_v f_v \mu_{eg}^{vv_2} \mu_{ge}^{v_1 v} \tag{11}$$

where the "frequency factor" $f_v$ is

$$f_v = \frac{2}{h} \frac{v_{eg}^{vv_1}}{v_{eg}^{vv_1 2} - v_0^2} \tag{12}$$

While in Placzek's polarizability theory conditions are presupposed which allow the (averaged) denominators in equation (9) to be extracted from the sums over $v$ which thereupon are easily evaluated (Placzek,[11] p. 267 *ff*), the conditions for the preresonance RE are more restricted. Particularly, the condition $v_{eg}^{vv_1} - v_0 \gg |v_{gg}^{v_1 v_2}|$ is no longer valid, and it is no longer possible to neglect the dependence of the denominators in equation (9), i.e., of the frequency factor $f_v$ in equation (11), on $v$. Shorygin's way of treating the problem is equivalent to the following semi-classical procedure. (A somewhat different representation of Shorygin's method is given by Behringer and Brandmüller,[17] p. 650 *ff*.)

Consider only one (nondegenerate) normal vibration in states $g$ and $e$ and for the respective normal coordinate (see Fig. 13) assume that

$$\bar{Q} = Q - b \tag{13}$$

with sufficiently large $b$ (distance of potential minima). Let $u_{gv}(Q)$ and $u_{ev}(\bar{Q})$ be the vibrational eigenfunctions in $g$ and $e$, respectively. Then, using the Born–Oppenheimer separation of electronic and vibrational eigenfunctions, we have, e.g.,

$$\mu_{eg}^{vv_2} = \int_{-\infty}^{+\infty} u_{ev}(Q - b) \mu_{eg}(Q) u_{gv_2}(Q) \, dQ \tag{14}$$

where $\mu_{eg}(Q)$ is the pure electronic transition moment which is parametrically dependent on the nuclear configuration $Q$. Now, if the two potential curves $g$ and $e$ are considered fixed relative to one

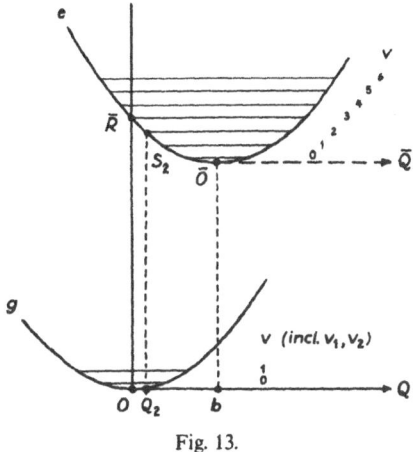

Fig. 13.

another, every $v$ of $e$ may be assigned to a certain value of $Q$ (e.g., $v = 2$ to $Q = Q_2$) by means of the point of intersection ($S_2$) near $\bar{R}$ of vibrational level $e, v$ with potential curve $e$. (This point of intersection approximately corresponds to the principal terminal maximum of the upper vibrational eigenfunction which by the Franck–Condon principle for small $v_1, v_2$ chiefly contributes to equation (11) and to which, for this step of the argument, the vibrational eigenfunction is reduced.) Thus, all sequences of discrete values labeled by $v$ (e.g., $v_{eg}^{vv_1}$ or $f_v$) may be translated into functions of $Q$ (e.g., $v_{eg}^{v_1}(Q)$ or $f(Q)$) which thereupon, if quantization is neglected, will be understood to be defined not only for the discrete $Q_i$, but also continuously for all $Q$. This procedure is applied to equation (11) which, by means of the closure property,

$$\sum_v u_{ev}(Q - b)u_{ev}(Q' - b) = \delta(Q - Q')$$

gives

$$\alpha_{gg}^{v_1v_2} = \sum_v f_v \int_{-\infty}^{+\infty} u_{ev}(Q - b)\mu_{eg}(Q)u_{gv_2}(Q)dQ$$

$$\times \int_{-\infty}^{+\infty} u_{ev}(Q' - b)\mu_{ge}(Q')u_{gv_1}(Q')dQ'$$

$$\approx \iint f(Q)\delta(Q - Q')u_{gv_2}(Q)\mu_{eg}(Q)\mu_{ge}(Q')u_{gv_1}(Q')dQdQ'$$

$$= (gv_2|f(Q)\mu_{eg}(Q)\mu_{ge}(Q)|gv_1) \tag{15}$$

in conventional simplified notation. Expansion of the operator in equation (15) into a Taylor's series in $Q$ at $Q = 0$ and restriction to the case of the Stokes R fundamental with $v_1 = 0$, $v_2 = 1$, and the linear term of the expansion and with

$$(g1|Q|g0) = \sqrt{\frac{\hbar}{2\omega}} = Q_0 \tag{16}$$

(where $\omega = 2\pi v = 2\pi |v_{gg}^{01}|$ is the angular frequency of vibration and $Q_0$ is the zero-point amplitude) yields

$$\alpha_{gg}^{01} = \left\{ \frac{\partial}{\partial Q} [f(Q)\mu_{eg}(Q)\mu_{ge}(Q)] \right\}_0 Q_0$$

$$= \mu_{eg}(0)\mu_{ge}(0) \left( \frac{\partial f}{\partial Q} \right)_0 Q_0 + f(0) \left\{ \frac{\partial}{\partial Q} [\mu_{eg}(Q)\mu_{ge}(Q)] \right\}_0 Q_0 \tag{17}$$

Here, the second term exactly represents Placzek's polarizability theory, while the first is characteristic of Shorygin's semi-classical treatment of the preresonance RE.

**Calculation of Effective Absorption Frequencies.** The two terms in equation (17) show different $v_0$-dependence. Use of the abbreviations

$$v_{eg}^{v_1}(0) \equiv v_{eg}$$

and

$$\left[ \frac{\partial v_{eg}^{v_1}(Q)}{\partial Q} \right]_0 \equiv v'_{eg}$$

yields

$$f(0) = \frac{2}{h} \frac{v_{eg}}{v_{eg}^2 - v_0^2} \tag{18}$$

and

$$\left( \frac{\partial f}{\partial Q} \right)_0 = -\frac{2}{h} \frac{v_{eg}^2 + v_0^2}{(v_{eg}^2 - v_0^2)^2} v'_{eg} \tag{19}$$

Provided that $v'_{eg} \neq 0$, expression (19) will increase faster than (18) when $v_0 \to v_{eg}$. Therefore, in the preresonance RE, the first term of equation (17) becomes prevalent over the second. This may be checked experimentally by comparison of the measured intensity dependence on $v_0$ with the dependence calculated separately by using equations (18) or (19), i.e.,

$$I_{mn} \propto (v_0 - v)^4/(v_{eg}^2 - v_0^2)^2 \qquad (20)$$

or

$$I_{mn} \propto (v_0 - v)^4 \frac{(v_{eg}^2 + v_0^2)^2}{(v_{eg}^2 - v_0^2)^4} \qquad (21)$$

where, by theory, $v_{eg}$ should approximately correspond to the long-wave vibronic absorption maximum.

For the practical realization of this comparison, it is easiest to determine experimentally the intensities $I_{mn}^{(i)}$ for (at least) two different exciting frequencies $v_0^{(i)}$ and then to calculate $v_{eg}$ from equations (20) or (21). An explicit description of this method inaugurated by Finkelstein[30,31] including certain modifications is given by Behringer.[17,19] The calculated $v_{eg}$ are called "effective absorption frequencies." This term suggests the possibility that there might be absorption bands active and nonactive for the RRE.*

**Comparison with True Absorption Frequencies.** The comparison of effective absorption frequencies calculated from RR spectra and true absorption frequencies taken from ultraviolet and visible absorption spectroscopy measurements shows that equation (21) generally leads to much better agreement than equation (20). This confirms the necessity of supplementing the polarizability theory term by the first term in equation (17).

Table VI presents the effective and true absorption wave numbers of a series of compounds after Finkelstein.[30] Calculations were made

---

* Kondilenko, Korotkov, and Strizhevsky[45-47] derived equations similar to (20) and (21), but with $(v_0 - v)^4$ replaced by $(v_0 - v)^2/v_0^2$, by neglecting the so-called Waller term in the exact quantum-mechanical dispersion equation leading to equation (7) (see, e.g., Heitler,[3] p. 192, and Wentzel,[12] p. 779). The utility of this step appears doubtful to this author for the following reasons: In order to preserve the relative simplicity of the equations, an operator different from the electric dipole-moment operator and less perspicuous has to be used. For Rayleigh scattering, the change may not be effected at all, the Waller term being essential in this case. When rigorous resonance conditions are approached, approximately $v_0^2(v_0 - v)^2 \rightarrow v_{eg}^4$ and the difference between the old and modified equations (both reduced to the resonance term) practically disappears. The experimental material presented by Tsenter and Bobovich[86] is not conclusive in demonstrating the advantage of the modified equations, for it is always possible to choose effective absorption frequencies such that for a restricted range of exciting frequencies the theoretical and measured intensity dependences are adjusted to one another. For nitrobenzene derivatives, at least in some cases, Shorygin's equation gives better agreement between true and effective absorption frequencies.

## TABLE VI
### Effective and True Absorption Wave Numbers After Finkelstein[30]

| Compound | Raman band wave number $\tilde{\nu}(cm^{-1})$ | Absorption band wave number | |
|---|---|---|---|
| | | Effective $\tilde{\nu}_{eff}(cm^{-1})$ | True $\tilde{\nu}_{abs}(cm^{-1})$ |
| Hexene-1 | 1642 | 50,000 | 58,000 |
| Cyclohexadiene-1,3 | 1576 | 45,000 | 50,000 |
| Alloocimene | 1626 | 40,000 | 37,000 |
| Nitropropane | 1380 | 50,000 | 45,500 |
| Nitrobenzene | 1348 | 41,000 | 39,000 |
| p-Nitrophenol | 1343 | 33,000 | 32,000 |

with equation (21), neglecting the R shift $\nu$ and using measurements with four different excitation lines (Hg c, e, k, q).

Table VII contains the results of similar measurements after Behringer,[19] p. 562 ff (cf., Shorygin and Osityanskaya[69,70]). The notation used in this table is as follows: $\tilde{\nu}$ is the R shift wave number of the band observed; $I_k$ and $I_e$ are its intensities excited by Hg k and Hg e, respectively; $\tilde{\nu}_{eff}$ is the effective wave number calculated by use of equation (21) with $\tilde{\nu}$ neglected; and $\tilde{\nu}_{abs}$ are the absorption band maxima taken from the literature.

## TABLE VII
### Effective and True Wave Numbers After Behringer[19]

| Compound | $\tilde{\nu}$ $(cm^{-1})$ | $I_k$ (IU) | $I_e$ (IU) | $\tilde{\nu}_{eff}$ $(cm^{-1})$ | $\tilde{\nu}_{abs}$ (Maxima) $(cm^{-1})$ |
|---|---|---|---|---|---|
| Alloocimene | 1629 | 13,750 | 5560 | 33,500 | 35,700 |
| Benzonitrile | 2225 | 1,160* | 447 | 32,900 | 37,000 |
| Hexene-1 | 1638 | 93 | 38 | 33,600 | >41,700 |
| Nitrobenzene | 1345 | 1,800 | 698 | 33,100 | 39,700 |
| Benzophenone | 1596 | 2,200 | 421 | 28,600 | 29,000, 41,000 |
| Coumarine | 1176 | 1,290 | 650 | 38,100 | 31,600, 38,000 |
| Diphenyl | 1590 + 1610 | 1,327 | 766 | 43,800 | 40,500 |
| Diphenyleneoxide | 1632 | 1,970 | 895 | 35,500 | 30,000, 35,500 |
| Fluorene | 1610 | 2,940 | 1040 | 36,800 | 34,000, 38,000 |
| cis-Stilbene | 1634 / 998 | 13,170 / 1,820 | 7260 / 950 | 41,200 } / 39,100 } | 36,000, 45,000 |

* Error of the intensity values is $\pm 20\%$, with the exception of the value marked by an asterisk where it is $\pm 50\%$.

The agreement is qualitatively satisfactory, but it should not be forgotten that several other examples have been found (Behringer,[19] p. 563) where neither equation (20) nor equation (21) leads to reasonable values for $\tilde{v}_{eff}$. The unsatisfactory results may be improved by taking account of $v$ in the above equations and of the proper $v_0$-dependence of the reference line intensities, but even then an exact identification of the hypothetical R-active absorption band is not possible in every case. The explanation for this has to be sought in the rather gross simplifications of the theory basic to equation (21) which cannot be expected to be justified for every molecule.

**Influence of Damping.** The influence of damping (neglected above) becomes remarkable when $|v_{eg} - v_0| < \delta_e^v$. The term $\delta_e^v$ may be estimated roughly from the half-value width (by theory equal to $2\delta_e^v$) of the vibrational bands appearing in the electronic absorption spectrum. In Table VIII after Shorygin and Krushinsky,[71] the relative deviations from $I_{NO_2, v(\pi)}/v_e^4$ for two pairs of exciting lines (Hg c, e in columns 3–5 and Hg e, k in columns 6–8, respectively) are listed for a series of para-substituted nitrobenzenes with increasing long-wave absorption (column 9). The agreement between measured and calculated values in general is much better if equation (21) is used instead of equation (20), except in those cases where $\tilde{v}_{eg} - \tilde{v}_0$ is small (cf., wave numbers of Hg c, e, k with $\tilde{v}_{eg}$). When the exciting frequency approaches close to the absorption band maximum, evidently neither equation (20) nor equation (21) remains valid. The intensity increase diminishes and, with excitation at the absorption peak, even stops. In this region, Shorygin and Krushinsky observed an approximate proportionality between R line intensity and absorption coefficient.[71] We shall discuss this proportionality below.

Figure 14 shows the absorption spectrum of p-nitrodimethylaniline in methanol,[18] from which $\delta_e^v$ may be estimated to be rather large (no vibrational structure visible), but not exceeding $d/2 \approx 2000\ cm^{-1}$. The absorption spectra of the other compounds being similar, this is about the order of magnitude for $\delta_e^v$ to be expected from the discrepancies in columns 3 and 5 or 6 and 8 of Table VIII according to equation (9) with $\delta_e^v \neq 0$.

We now give an explanation for the above-mentioned approximate proportionality between R line intensity and absorption coefficient for excitation near the resonance maximum, presuming $\delta_e^v = \delta_e$ is approximately independent of $v$ and considerably larger than the

## TABLE VIII
### Relative Deviations from the $\nu_e^4$-Law After Shorygin and Krushinsky[71]

| Compound | Solvent | Excitation by Hg c, e ($\bar{\nu}_0^{(1)} = \bar{\nu}_c$, $\bar{\nu}_0^{(2)} = \bar{\nu}_e$)* | | | Excitation by Hg e, k ($\bar{\nu}_0^{(1)} = \bar{\nu}_e$, $\bar{\nu}_0^{(2)} = \bar{\nu}_k$)* | | | Absorption band maximum $\bar{\nu}_{eg}$ (cm$^{-1}$) |
|---|---|---|---|---|---|---|---|---|
| | | Exp.† | Calculated with equation | | Exp.† | Calculated with equation | | |
| | | | (20)‡ | (21)§ | | (20)‡ | (21)§ | |
| p-Nitrotoluene | Cyclohexane | 2.0 | 1.4 | 2.1 | 1.6 | 1.2 | 1.5 | 38,000 |
| p-Nitroanisol | Methanol | 2.7 | 1.6 | 2.8 | 1.5 | 1.3 | 1.7 | 34,200 |
| p-Nitroaniline | Methanol | 11 | 3.8 | 14 | 7.3 | 2.9 | 8.5 | 27,000 |
| p-Nitrodimethylaniline | Methanol | 38 | 5.8 | 34 | 2.2 | 7.0 | 50 | 25,700 |
| p-Nitrodiethylaniline | Methanol | 46 | 8.4 | 71 | 1.8 | 44 | 1960 | 25,000 |

* $\bar{\nu}_c = 18,307$ cm$^{-1}$, $\bar{\nu}_e = 22,938$ cm$^{-1}$, and, $\bar{\nu}_k = 24,705$ cm$^{-1}$.

† $\dfrac{I_{NO_2}^{(1)}}{I_{NO_2}^{(2)}} \cdot \dfrac{(\bar{\nu}_0^{(2)} - \bar{\nu}_{NO_2})^4}{(\bar{\nu}_0^{(1)} - \bar{\nu}_{NO_2})^4}$.

‡ $\left(\dfrac{\bar{\nu}_{eg}^2 - \bar{\nu}_0^{(1)2}}{\bar{\nu}_{eg}^2 - \bar{\nu}_0^{(2)2}}\right)^2$.

§ $\left(\dfrac{\bar{\nu}_{eg}^2 - \bar{\nu}_0^{(1)2}}{\bar{\nu}_{eg}^2 - \bar{\nu}_0^{(2)2}}\right)^4$.

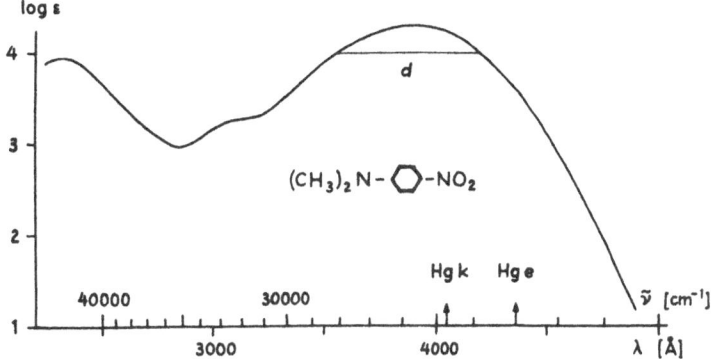

Fig. 14. Absorption spectrum of p-nitrodimethylaniline in methanol solution.[18]

spacing of vibrational levels $v$. Maintaining the other qualifications made after equation (9), we can now (by a procedure analogous to that in Placzek's polarizability theory) simplify this equation for $v_1 = 0$ and $v_2 = 1$ (fundamental vibration $Q$) to

$$\alpha_{gg}^{01} = \frac{1}{h} \frac{\sum\limits_{v} \mu_{eg}^{v1} \mu_{ge}^{0v}}{v_{eg} - v_0 + i\delta_e} \tag{22}$$

where $v_{eg}$ as the average value for the $v_{eg}^{v0}$ represents the absorption band maximum determined by the Franck–Condon transition from $v_1 = 0$. Evaluation of the sum gives to first order in $Q$

$$\sum_{v} \mu_{eg}^{v1} \mu_{ge}^{0v} = \left\{ \frac{\partial}{\partial Q} [\mu_{eg}(Q) \mu_{ge}(Q)] \right\}_0 Q_0 = (\mu\mu' + \mu'\mu) Q_0 \tag{23}$$

with the abbreviations

$$\mu = \mu_{eg}(0) \quad \text{and} \quad \mu' = \left[ \frac{\partial}{\partial Q} \mu_{eg}(Q) \right]_0$$

Then by equation (7) (cf., the footnote on p. 180) the intensity scattered on the average by one molecule into $4\pi$ is

$$I_{gg}^{01} = \frac{2^7 \pi^5}{3^2 h^2 c^4} I_0 (v_0 - v)^4 2 Q_0^2 \frac{(\mu \cdot \mu)(\mu' \cdot \mu') + (\mu \cdot \mu')^2}{(v_{eg} - v_0)^2 + \delta_e^2} \tag{24}$$

On the other hand, the absorption coefficient in the resonance region in zeroth order is given by (cf., Heitler,[3] pp. 178, 186)

$$\chi(\nu_0) = \frac{2^2\pi}{3hc}Z\delta_e\nu_0\frac{\mathbf{\mu}\cdot\mathbf{\mu}}{(\nu_{eg} - \nu_0)^2 + \delta_e^2} \tag{25}$$

(where $Z$ number of molecules in unit volume); in the case of pure radiation damping

$$\delta_e = \frac{2^5\pi^4}{3hc^3}\nu_0^3\mathbf{\mu}\cdot\mathbf{\mu} \tag{26}$$

equation (25) may be written as

$$\chi(\nu_0) = \frac{2^7\pi^5}{3^2h^2c^4}Z\nu_0^4\frac{(\mathbf{\mu}\cdot\mathbf{\mu})^2}{(\nu_{eg} - \nu_0)^2 + \delta_e^2} \tag{27}$$

The approximate proportionality of $I_{gg}^{01}$ and $\chi(\nu_0)$ is obvious. [For higher-order approximations of equation (25) and necessarily in the case of vanishing $\mathbf{\mu}\cdot\mathbf{\mu}$, Albrecht's theory[14,15] of "forbidden intensities" in vibronic spectra might prove useful.*]

### Selection of Raman Bands in Preresonance Spectra

**Raman Intensities and Structure of Vibronic Spectra.** By comparison of R spectra with ultraviolet and visible vibronic absorption and fluorescence spectra of numerous substances, it was empirically established[21] that in R spectra, especially under preresonance excitation, those vibrations which determine the vibrational structure of the absorption band nearest the exciting line appear with high intensities. When fluorescence emission takes place, the vibrational structure of the fluorescence band with regard to selection of vibrations often agrees with the vibrational structure of the first long-wave electronic absorption band and, therefore, also matches the most intensive R lines. The vibration frequency values abstracted from this fluorescence vibrational structure agree even better with R frequencies than those from absorption vibrational structure.

---

* A similar treatment of this problem was indicated by Shorygin and Krushinsky;[78,79] for a more elementary discussion, see Behringer.[20] Sushchinsky and Zubov[83] tried to show by exact quantum-mechanical calculations and experimental investigations on carbon tetrachloride, benzene, and some polyenes that the proportionality between R line intensity and absorption coefficient is a generally valid law. It seems to this author that their final equation contains an erroneous factor $k_\sigma^4$ (corresponding to $h^4\nu_e^4$). When this factor is omitted, the agreement between their theory and experimental measurements is much better, but still not fully satisfactory.

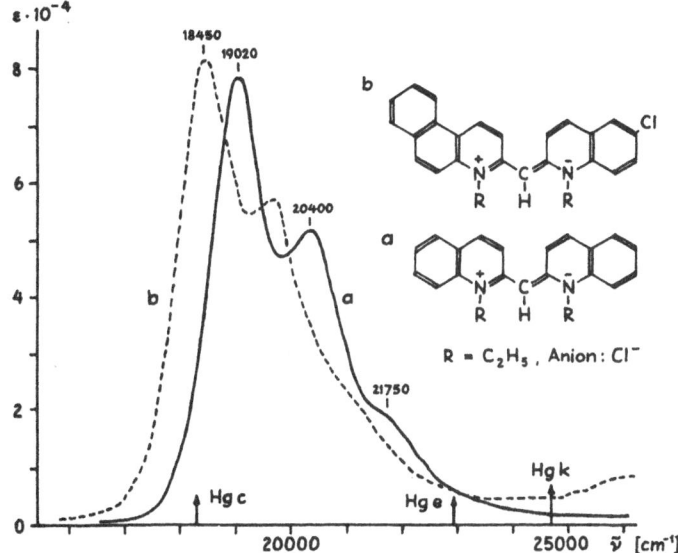

Fig. 15. Absorption spectra of pseudo-isocyanine (a) and 5,6-benzo-6'-chloro-pseudo-isocyanine (b) after Maier and Dörr.[49,50] Monomers (chlorides) in methanol solutions.

Examples for these rules are $\beta$-carotene (Figs. 7 and 12) with the R line 1521 cm$^{-1}$ approximately corresponding to the average distance 1540 cm$^{-1}$ of vibrational structure maxima and the pseudo-isocyanines mentioned earlier (Fig. 8) with the R lines 1365 or 1358 cm$^{-1}$, respectively, approximately matching the absorption peaks' distance 1350 cm$^{-1}$ (Fig. 15). Several other examples are found in the work of Behringer and Brandmüller[21] and will be given in a later section entitled "Rigorous Resonance Raman Effect."

It seems justifiable to conclude from these empirical findings that in preresonance RE always (even when the vibrational structure of the absorption and fluorescence band is latent because of broadening or when several absorption bands overlap) those vibrations which characterize the vibrational structure of the electronic band nearest the exciting frequency are particularly intensified. This conclusion also is consistent with theoretical predictions.

**Unequal Intensity Variations of Different Raman Lines.** When the exciting frequency approaches the region of electronic absorption, the intensity increase of different R lines of the same molecule generally appears to be nonuniform. The calculation of effective absorption frequencies, therefore, necessarily leads to different values for the individual R lines.

For evidence, see Table IX which contains the effective absorption frequencies for carbon tetrachloride calculated by Hofmann and Moser[37] from Fig. 9 by means of equation (21) with damping neglected. The calculated values of $v_{eff}$ for all lines approximately fall into the region of long-wave electronic absorption which monotonically increases in intensity up to the measuring limit at 1500 Å, but these values differ much among themselves.

Unfortunately, sufficient exact intensity measurements made on several R lines of an individual substance using different exciting frequencies and keeping all other operating conditions unchanged are very scarce. This indeed would be the only way of getting a true picture of the RRE free of falsifications by additional and often uncontrollable effects. In spite of this scarcity of reliable experimental material, the disparity of intensity variation among different R lines of the same compound as a general rule seems to be beyond doubt. For additional evidence, the reader is referred to Table II and the work of Sushchinsky and Zubov.[83]

Before presenting arguments to explain these dissimilar intensity variations, we add some results found by the less reliable method of exchanging compounds or of varying solvents or other parameters in a series of measurements, simultaneously keeping the exciting frequency constant. The conclusions drawn from these latter investigations have to be accepted with caution, since it is by no means always certain that by variation of the parameter in question the vibronic absorption band

## TABLE IX
**Effective Absorption Wave Numbers and Wavelengths for Carbon Tetrachloride After Hofmann and Moser[37]**

| | Raman band $\tilde{v}$ (cm$^{-1}$) | | | |
|---|---|---|---|---|
| | 217 | 313 | 458 | 760–791 |
| $\tilde{v}_{eff}$ (cm$^{-1}$) | 80,000 | 71,000 | 65,000 | 57,500 |
| | $\pm 5,500$ | $\pm 4,000$ | $\pm 3,000$ | $\pm 1,500$ |
| $\lambda_{eff}$ (Å) | 1,250 | 1,430 | 1,550 | 1,750 |
| | $\pm 100$ | $\pm 100$ | $\pm 100$ | $\pm 50$ |

## TABLE X
### Intensity Growth of $a_1$ Raman Bands $v_1$ (Aromatic Ring, Totally Symmetric) and $v_2$ (Nitro Group, Symmetric) of para-Substituted Nitrobenzene Derivatives[19]

| Compound | Solvent | $\bar{v}_1$ (cm$^{-1}$) | $I_{v_1}$ (IU) | $I_{v_1}$* | $\bar{v}_2$ (cm$^{-1}$) | $I_{v_2}$ (IU) | $I_{v_2}$* |
|---|---|---|---|---|---|---|---|
| Nitrobenzene | CCl$_4$ | 852 | 53† | 7.6 | 1345 | 698† | 100 |
| p-Chloronitrobenzene | CCl$_4$ | 857 | 65† | 4.2 | 1338 | 1,370† | 100 |
| p-Nitrotoluene | CCl$_4$ | 856 | 135‡ | 9.2 | 1337 | 1,470‡ | 100 |
| p-Nitrophenetole | CCl$_4$ | 858 | 240† | 10.5 | 1341 | 2,290† | 100 |
| p-Nitrophenol | CH$_3$OH | 368 | 800§ | 14.8 | 1336 | 5,400§ | 100 |
| p-Nitroaniline | C$_6$H$_6$ | — | — | — | 1334 | 17,400§ | 100 |
| p-Nitrodimethylaniline | C$_6$H$_6$ | 355 | 90,000‡ | 22.5 | 1316 | 400,000‡ | 100 |
| p-Nitrodimethylaniline | CH$_3$OH | 354 | 1,060,000‡ | 85 | 1300 | 1,190,000‡ | 100 |
| K-p-Nitrophenolate | CH$_3$OH | 854 | 1,170,000‡ | 317 | 1279 | 370,000‡ | 100 |

$I_{v_1}$* relative intensity referred to $I_{v_2}$* normalized to 100.

† Error, ±5–10%.
‡ Error, ±20–30%.
§ Error, ±50%.

essentially suffers only a frequency shift and structurally remains constant. However, if this objection is neglected, we can state the following: Tables IV and V show that, contrary to what might be expected for carbon tetrachloride, the R wave number cannot be considered as the only decisive factor for the intensity rate of change. The intensity variations are quite irregular with respect to R wave number. Furthermore, while the data collected in Table I might suggest that nontotally symmetric or degenerate vibrations are more enhanced than the totally symmetric, counterevidence is given by many substances, e.g., the nitrobenzene derivatives.[21,80] Often vibration lines of the same symmetry type behave differently. Table X presents intensity values (excited by Hg e) of two $a_1$-bands of nitrobenzene derivatives.[19] By approach to resonance, the intensity of the

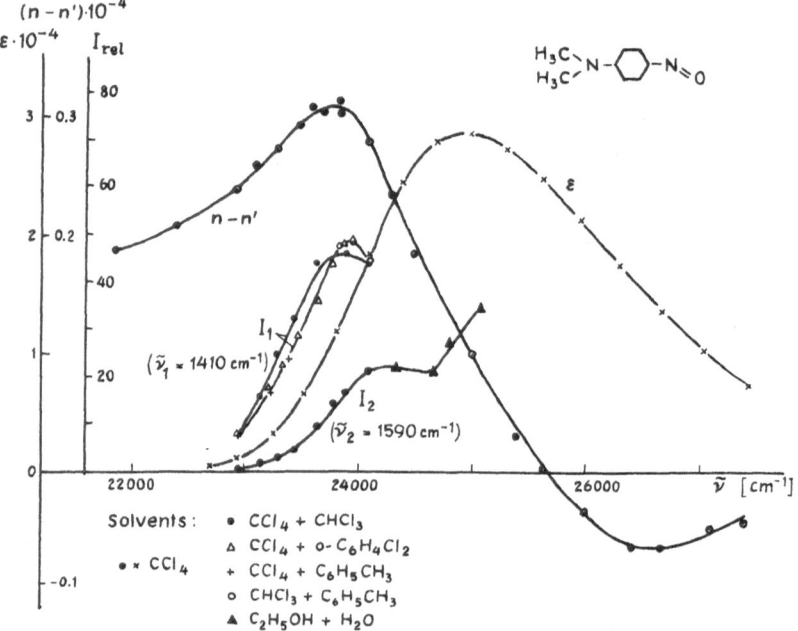

Fig. 16. $\nu_0$-Dependence of R intensities $I_{rel}$ (in relative units 'referred to an arbitrary intensity unit) of p-nitrosodimethylaniline after Tsenter and Bobovich.[87] Excitation by Hg e. $\varepsilon$ is the absorption coefficient for the carbon tetrachloride solution. The absorption curves for solutions in other solvents are similar in shape, but shifted along the $\nu$-axis. If it is assumed that this is equivalent to shifting of $\nu_0$, Raman intensities are plotted with identical shift so that continuous intensity curves $I_1$ and $I_2$ result. $n$ is the refractive index of the solution, $n'$ that of the solvent. Intensity error, 10–20%.

totally symmetric aromatic-ring vibration $\tilde{\nu}_1 \approx 850 \text{ cm}^{-1}$ ($\omega_4$ in Kohlrausch's[5] notation) is more enhanced than that of the symmetric $NO_2$ vibration $\tilde{\nu}_2 \approx 1300\text{–}1350 \text{ cm}^{-1}$.

Figure 16 shows the $\nu_0$-dependence of p-nitrosodimethylaniline R line intensities $I_1$ and $I_2$ referred to $\tilde{\nu}_1 = 1410 \text{ cm}^{-1}$ (coupled nitroso-group and benzene-ring vibration) and $\tilde{\nu}_2 = 1590 \text{ cm}^{-1}$ (benzene-ring vibration), together with the absorption coefficient $\chi(\nu_0)$ and the refraction index after Tsenter and Bobovich.[87] A surprising feature is the fact that as $\nu_0$ approaches the resonance maximum $I_2$ decreases and then increases again.

**Reasons for Dissimilarity of Intensity Variations.** At present it is not yet possible to explain satisfactorily in concrete cases the dissimilar intensity variations of individual R lines when excitation approaches resonance. However, some comment may be given.

*Different Raman-Active Absorption Bands.* For every molecule, the existence of several electronic absorption bands which do not necessarily agree in their vibrational structures nor overlap at identical frequencies is certain from absorption spectroscopy. Therefore, by the rule previously stated in the section entitled "Raman Intensities and Structure of Vibronic Spectra," it is possible that the individual molecular vibrations correspond to R-active electronic absorption bands at different distances from the exciting frequency given.

Provided that the ultraviolet absorption spectrum of the nitro-benzene derivatives (Table X) at least in the long-wave parts is little enough affected by chemical substitutions, the simplest explanation for the discussed behavior of R intensities would be to assume that the $NO_2$ group causes absorption (with $\tilde{\nu}_2$-vibration structure) at wavelengths somewhat blue-shifted relative to the first aromatic-ring absorption band (in benzene, the so-called $L_b$-band[44] at $\sim 2600 \text{ Å}$ with distinct $\tilde{\nu}_1$-vibration structure). This would be in accord with arguments of the theory of quantum-mechanical resonance for nitro-benzene (cf., Wheland,[8] p. 287, and Behringer,[19] p. 564).

*Influence of Vibrational Structure.* As will be demonstrated later by examples for the rigorous RRE, it is not only the frequency site, intensity, and width of the electronic absorption band as a whole, but also the corresponding properties of the superimposed individual vibration bands relative to the incident frequency that influence RR intensities. A detailed examination of this statement in the preresonance RE meets with the extreme difficulty not only of resolving the vibrational structure in the exterior parts of the absorption band, but also of measuring R intensities with the high accuracy demanded.

Perhaps the example given in Fig. 16 could be quoted here, the course of $I_2$ suggesting some relation to the vibrational structure, but it seems more probable to this author that in this case the molecular structure of the solute is gradually changed by solvent effects, as has been suggested by Maier.[49,50]

From a theoretical point of view [see equations (17) and (24)] $\mu = \mu_{eg}(0)$ as well as $\mu' = (\partial\mu_{eg}(Q)/\partial Q)_0$ determine R intensities. Although in the preresonance RE the first term in equation (17) connected with pure electronic absorption is generally more important, very far from and (provided $\delta_e$ is large) very close to resonance the second term containing $\mu'$ must be taken into consideration. Both terms of equation (17) taken together with $\mu'$ depending on the type of vibration might influence the course of intensity variation. Note that $(\partial f/\partial Q)_0$ depends on the inclination in direction $Q$ of the tangent plane in $\bar{R}$ on the potential surface (Fig. 13) and, therefore, on the relative deformation of nuclear equilibrium configurations in ground and excited states (quantity $b$ in Fig. 13).

*Degeneracy of the Intermediate State.* There is another aspect to be considered: Table I shows a remarkable favoring of degenerate vibrations. Although this is certainly not the rule for all molecules, it might be worthwhile (by modification of the semi-classical theory) to develop a detailed theory of the RRE for molecules with degenerate intermediate electronic states. Degeneracy of the intermediate state is requisite for degenerate vibrations to be R-active in molecules with totally symmetric ground states. Consider, for example, benzene (Fig. 17), for which unfortunately no sufficient RR data are available (*cf.*, Sushchinsky and Zubov,[83]). ($T_x$ means $x$-polarization of dipole-transition moment; for simplicity the pure electronic states are drawn for initial and intermediate states $m$ and $r$.) According to the theory of

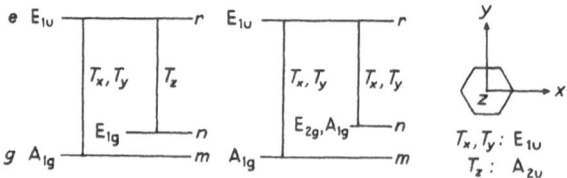

Fig. 17. Electronically allowed transitions in benzene.

electronic spectra (see, e.g., Sandorffy,[7] p. 98, and Sponer and Teller,[82] p. 148) the lowest excited electronic states of benzene have in succession the symmetries $B_{2u}$, $B_{1u}$, and $E_{1u}$. Pure electronic transition is allowed only for $E_{1u}$; its transition moment is polarized in the ring plane $(x, y)$. In the RE, only $A_{1g}$, $E_{1g}$, and $E_{2g}$ are allowed; out of the three pure electronic states mentioned above, only $E_{1u}$ may serve as intermediate state. The figure shows the possible transitions in the RRE. Because of the Franck–Condon principle and the Jahn–Teller effect for state $E_{1u}$, in reality not the pure electronic state $E_{1u}$ but a definite superposition with an arbitrary number of totally symmetric and an even number of nontotally symmetric vibration quanta will serve as intermediate state. Perhaps by detailed investigations of RR intensities and by use of the argument given in the immediately preceding section a judgment may be formed on the actual deformation of nuclear equilibrium configurations in the R-active excited electronic state relative to the ground state.

*Concurrent Radiationless Transitions.* It is known by experience in all cases investigated so far that by continual exaltation of the exciting frequency the observation of the RRE becomes more and more difficult and ultimately impossible. The radiation energy absorbed will be instantly reemitted no longer, but will be consumed by other processes concurrent with scattering. These processes are always either identical or connected with radiationless transitions from the intermediate state reached in the initial (virtual) absorption act involved in scattering. It is plausible to assume that the probability for these radiationless transitions, in general, will increase faster (but in an individually different way) than that for scattering reemission. Consequently, RR intensities by these concurrent processes will gradually be diminished when $v_0$ grows. An example that demonstrates this possibility will be given in a later section.

**Depolarization Ratios and Polarization of Electronic Transitions.** When the excited electronic state $e$ is nondegenerate, equation (17) of the semi-classical theory is valid for each nondegenerate normal vibration $Q$ of a polyatomic molecule. In the vicinity of resonance, the second term may be approximately neglected. With use of the notation of polarizability theory

$\alpha'_{ik} = (\partial \alpha_{ik}/\partial Q)_0$ derived polarizability component

$i,k = x, y, z$      molecule fixed system

equation (17) for the preresonance RE then reads (with omission of the factor $Q_0$)

$$\alpha'_{ik} = \mu_{eg,i}(0)\mu_{ge,k}(0)\left(\frac{\partial f}{\partial Q}\right)_0 \qquad (28)$$

For natural (unpolarized) irradiation, the depolarization ratio of a R band is defined by (Herzberg,[4] p. 247, and Wilson et al.,[9] p. 51)

$$\rho = \frac{6\beta^2}{45\alpha^2 + 7\beta^2} \qquad (29)$$

where the *spherical part* $\alpha$ and the *anisotropy* $\beta$ of the derived polarizability are given by

$$\alpha = \frac{1}{3}\sum_i \alpha'_{ii}$$

$$\beta^2 = 9\alpha^2 - 3\sum_{\substack{i,k \\ i<k}} (\alpha'_{ii}\alpha'_{kk} - \alpha'_{ik}\alpha'_{ki}) \qquad (30)$$

As was proved by polarization spectroscopy in many organic molecules, the dipole-transition moments characterizing the single ultraviolet or visible absorption (and eventually fluorescence) bands have a well-defined molecule-fixed orientation. For example, as shown by Eckert and Kuhn[29] in $\beta$-carotene [Fig. 18(a)] the transition moments of the principal absorption band A at 4500–5000 Å (cf., Fig. 12) and the weaker C-band at 2900 Å are directed along the molecular axis $x$, while in 15,15'-cis-$\beta$-carotene [Fig. 18(b)] an additional weak band B

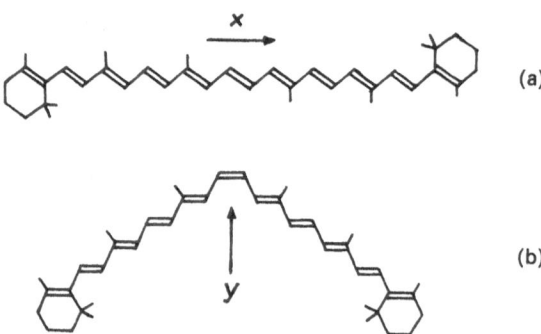

Fig. 18. (a) $\beta$-carotene. (b) 15,15'-cis-$\beta$-carotene. After Eckert and Kuhn.[29]

at 3400 Å ("cis-peak") appears with a transition moment in the traverse direction $y$. In benzene, the allowed electronic transition $A_{1g} \rightarrow E_{1u}$ leading to the so-called[44] $B_b$-band at 1650–1850 Å is polarized in the ring plane $(x, y)$ without definite direction. The situation is more complex for electronically forbidden transitions which are induced by vibrations, as is the case for the benzene $A_{1g} \rightarrow B_{2u}$ band $L_b$ at 2600 Å and the $A_{1g} \rightarrow B_{1u}$ band $L_a$ at 2000 Å.

When the electronic transition moment $\mu_{eg}(0)$ has a definite direction, e.g., $x$, all $\alpha'_{ik}$ except $\alpha'_{xx}$ vanish and by the above equations

$$\rho = \tfrac{1}{2} \qquad (31)$$

Equation (31) should hold for all nondegenerate fundamental vibrations in molecules with a practically (respective to the given excitation frequency) single R-active absorption band which is electronically allowed and which results from a transition lying in a definite molecule-fixed direction. (The latter condition presupposes nondegeneracy of the electronic states $g$ and $e$.) From the theorem (Herzberg,[4] p. 249; Wilson et al.,[9] p. 52; Placzek,[11] p. 291) that $\rho < \tfrac{6}{7}$ only for totally symmetric fundamentals, it may be concluded at once that under the above conditions only totally symmetric fundamentals are allowed in the preresonance RE.

Table XI shows for several nitro compounds depolarization ratios for the symmetric $NO_2$-valence vibration line $\tilde{v}_1$ and a line $\tilde{v}_2$ with a wave number similar to that of the antisymmetric $NO_2$ vibration $\tilde{v}(\sigma)$ as measured by Bobovich and Tsenter,[24,85] Rea,[58] and Schrötter[80] by excitation with Hg e. In the literature[19,24] $\tilde{v}_2$ is identified with $\tilde{v}(\sigma)$. In compounds with symmetry $\mathscr{C}_{2v}$ this would be a contradiction to the selection rule given above. Computational analysis[28] has shown, however, that this line is likely to be assigned to a CC-ring vibration of symmetry $A_1$ in $\mathscr{C}_{2v}$.

Depolarization ratios for the C=C-valence vibration after Bobovich and Tsenter[24] and Harrand[34] (excitation by Hg e) are given in Table XII.

Tables XI and XII show that (within experimental error) $\rho = 0.5$ indeed must be considered as the limiting value for the depolarization ratio of nondegenerate vibrations of characteristic groups such as $NO_2$ or C=C within molecules with linearly arranged $\pi$-bonding systems of sufficient length. It is known from absorption spectroscopy that such molecules possess a very intense first (long wave) absorption band of definite molecule-fixed polarization. When the linearity of the

## TABLE XI
### Depolarization Ratios for NO₂ Compounds

| Formula | Compound | Solvent | Symmetrical NO₂ vibration $\tilde{v}_1(cm^{-1})$ | $\rho$ | Vibration ~1510 cm⁻¹ $\tilde{v}_2(cm^{-1})$ | $\rho$ | Reference |
|---|---|---|---|---|---|---|---|
| $CH_3NO_2$ | Nitromethane | Pure | 1375 + 1400 | 0.38 | ? | ? | 85 |
| phenyl–$NO_2$ | Nitrobenzene | Acetone | 1345 | 0.28 | 1510 | 0.72 | 24 |
| phenyl (meta $NO_2$, $NO_2$) | m-Dinitrobenzene | Acetone | 1340 | 0.28 | 1530 | 0.97 | 24 |
| $H_3C$–phenyl–$NO_2$ | p-Nitrotoluene | CCl₄ | ? | 0.31 | ? | ? | 58 |
| | | Acetone | 1335 | 0.35 | ? | ? | 24 |
| $H_2N$–phenyl–$NO_2$ | p-Nitroaniline | Benzene | 1334 | 0.42 | 1505 | 0.6 | 80 |
| | | CCl₄ + acetone | 1325 | 0.50 | 1510 | 0.43 | 24 |
| $(CH_3)_2N$–phenyl–$NO_2$ | p-Nitrodimethylaniline | Benzene | 1318 | 0.45 | 1508 + 1522 | 0.5 | 80 |
| $(C_2H_5)_2N$–phenyl–$NO_2$ | p-Nitrodiethylaniline | Benzene | 1318 | 0.45 | 1508 + 1518 | 0.5 | 80 |
| $CH_3O$–phenyl–$CH{=}CH{-}NO_2$ | | CCl₄ | 1325 | 0.47 | 1500 | 0.40 | 24 |
| | | Acetone | 1325 | 0.49 | 1500 | 0.45 | 24 |
| $C_2H_5O$–phenyl–$NO_2$ | | Acetone | 1340 | 0.42 | 1500 | 0.50 | 24 |

## TABLE XII
### Depolarization Ratios for C=C Valence Vibration

| Formula | Compound | Solvent | $\tilde{v}(\mathrm{cm}^{-1})$ | $\rho$ | Reference |
|---|---|---|---|---|---|
| [benzene ring]—$CH=CH_2$ | Styrolene | Acetone | 1640 | 0.28 | 24 |
| [furan ring]—$CH=CH-NO_2$ | | $CCl_4$ | 1620 | 0.34 | 24 |
| [benzene ring]—$C(CH_3)=CH-CO-OC_2H_5$ | Ethyl-$\beta$-methylcinnamate | Pure | 1636 | 0.36 | 34 |
| $O_2N$—[benzene ring]—$CH=CH-NO_2$ | | $CCl_4$ + acetone | 1630 | 0.44 | 24 |
| [benzene ring]—$CH=CH$—[benzene ring] | Stilbene | Dichloroethane | 1620 | 0.49 | 24 |
| [benzene ring]—$CH=CH-CO-OC_2H_5$ | Ethylcinnamate | Pure | 1640 | 0.45 | 34 |
| [benzene ring]—$CH=CH-CHO$ | Cinnamaldehyde | Pure | 1628 | 0.53 | 34 |
| [furan ring]—$CH=CH-CH=CH-NO_2$ | | $\begin{cases} CCl_4 \\ Acetone \end{cases}$ | 1620<br>1620 | 0.48<br>0.50 | 24<br>24 |

## TABLE XIII

**Depolarization Ratios for Aromatic-Ring Vibrations $\omega_3 \approx 1000\ cm^{-1}$ and $\omega_8 \approx 1600\ cm^{-1}$**

| Formula | Compound | Solvent | $\omega_3(cm^{-1})$ | $\rho$ | $\omega_8(cm^{-1})$ | $\rho$ | Reference |
|---|---|---|---|---|---|---|---|
| ⬡ | Benzene | Pure | 992 | $0.06 \pm 0.01$ | $1584 + 1605$ | $0.84 \pm 0.02$ | 90 |
| ⬡—CH₃ | Toluene | Pure | 1010 | 0.03 | 1590 | 0.88 | 24 |
| ⬡(CH₃)₂ | m-Xylene | Pure | 1000 | 0.04 | 1590 | 0.93 | 24 |
| ⬡—C(=O)—CH₃ | Acetophenone | Pure (?) | 1000 | 0.07 | 1599 | 0.65 | 65 |
| ⬡—CH=CH₂ | Styrolene | Pure | 1007 | 0.17 | 1609 | 0.69 | 34 |
| ⬡—CH=CH—⬡ | Stilbene | Dichloroethane | 1000 | 0.28 | 1590 | 0.59 | 24 |
| ⬡—CH=CH—CHO | Cinnamaldehyde | Pure | 1007 | 0.26 | 1604 | 0.45 | 34 |

electronic oscillator within the molecule is perturbed or missing, as in the first compounds listed in the above tables, equation (31) no longer holds.

Table XIII presents depolarization ratios for the symmetric and antisymmetric aromatic-ring vibrations $\omega_3 \approx 1000 \, cm^{-1}$ (symmetry $A_1$) and $\omega_8 \approx 1600 \, cm^{-1}$ ($B_1$) (in Kohlrausch's[5] notation) in several aromatic compounds after Bobovich and Tsenter,[24] Harrand,[34] Shorygin,[65] and Wittek[90] (excitation by Hg e). Evidently, $\rho \to \frac{1}{2}$ for $\omega_8$ but not for $\omega_3$.

Perhaps the values $\rho = \frac{2}{9}, \frac{6}{7}, 0,$ and $\frac{8}{11}$ resulting from the following assumptions

$$\left. \begin{array}{l} \alpha'_{xx} = \alpha'_{yy} \\ \alpha'_{xx} = -\alpha'_{yy} \\ \alpha'_{xx} = \alpha'_{yy} = \alpha'_{zz} \\ \alpha'_{xx} = \alpha'_{yy} = \alpha'_{zz} \end{array} \right\} \text{ all other } \alpha'_{ik} = 0 \qquad (32)$$

respectively, also play the role of limiting values in these tables (cf., Rea[58] and Tsenter and Bobovich[85]). So far it has not been possible to give any satisfactory theoretical arguments or final experimental verification for this supposition, but it seems that proper consideration of degeneracy of the intermediate state, of vibration symmetry, and of electronically forbidden transitions (e.g., in first absorption bands of benzene) might yield some elucidation of this interesting problem. We note that the dissection of $\alpha'_{ik}$ into a $\pi$- and a $\sigma$-electron part proposed by Savin and Sobelman[59-61] (metallic model of molecule) and in some respect successfully used for explanation by Zubov[91] (see below) shows some analogy with equation (17).

Tsenter and Bobovich[85] presented $\rho$-measurements on compounds with essentially single or linearly arranged bonding systems using Hg c and Hg e for excitation. Within experimental accuracy, no change of $\rho$ was observed. This was regarded as evidence that $\rho$ approaching 0.5 has nothing to do with the RRE. This conclusion was later disproved by Zubov.[91] The results of his depolarization measurements by use of several visible and ultraviolet Hg exciting lines are reproduced in Table XIV.

## RIGOROUS RESONANCE RAMAN EFFECT

### Irradiation on the Short-Wave Side of Absorption bands

In most substances examined, the observation of the RRE is impossible when an exciting frequency that lies in the short-wave

TABLE XIV

Dependence of Depolarization Ratio on Exciting Frequency After Zubov[91]

| Compound | $\bar{\nu}$ (cm$^{-1}$) | Depolarization ratio $\rho$ by excitation with Hg line (Å) | | | | |
|---|---|---|---|---|---|---|
| | | 5461 | 4358 | 3132 | 3126 | 3021 |
| Carbon tetrachloride | 217 | 0.86 | 0.86 | 0.86 | 0.86 | 0.86 |
| | 313 | 0.86 | 0.86 | 0.86 | 0.86 | 0.86 |
| | 459 | 0.04 | 0.04 | 0.04 | 0.04 | 0.04 |
| Benzene | 992 | 0.05 ± 0.01 | 0.06 ± 0.01 | 0.08 ± 0.02 | 0.07 ± 0.03 | — |
| Toluene | 1004 | 0.04 ± 0.01 | 0.04 ± 0.02 | 0.09 ± 0.02 | 0.07 ± 0.02 | — |
| Pentene-1 | 1642 | — | 0.12 | 0.12 ± 0.05 | — | 0.17 ± 0.03 |
| Pentadiene-1,3 | 1655 | 0.29 ± 0.06 | 0.31 ± 0.02 | 0.43 ± 0.02 | 0.44 ± 0.04 | 0.48 ± 0.07 |
| 2-Methylbutadiene-1,3 | 1638 | 0.23 ± 0.02 | 0.21 ± 0.02 | 0.38 ± 0.02 | 0.39 ± 0.01 | 0.53 ± 0.04 |
| 1,2-Disilylethane | 560 | — | 0.49 ± 0.04 | 0.43 ± 0.04 | 0.44 ± 0.02 | 0.40 ± 0.05 |
| | 2157 | 0.17 ± 0.02 | 0.17 ± 0.01 | 0.17 ± 0.03 | 0.16 ± 0.03 | 0.22 ± 0.04 |

region beyond the maximum of the first absorption band is chosen. Therefore, for example, $\beta$-carotene and the pseudo-isocyanines did not yield a RR spectrum when excited by Hg e or Hg k.[19,49] The probable reason for this phenomenon has already been discussed in the section entitled "Concurrent Radiationless Transitions."

However, in some substances the attempt to observe R spectra by irradiation on the short-wave side of the first absorption band was successful. For instance, a RRE of a methanol solution of malachite green (first absorption maximum at 6200 Å) by excitation with Hg e has been reported.[18,19] Shorygin and Ivanova[72] succeeded in observing RR spectra of 4-nitro-4'-dimethylaminostilbene (trans) in pyridine solution by excitation with Hg c, e, and k. Figure 19 shows the positions of the exciting lines (solid arrows) and of the R lines assigned to the $NO_2$ vibration $\tilde{\nu}(\pi) = 1340 \text{ cm}^{-1}$ (broken arrows, length proportional to intensity value) relative to the absorption band with its maximum at $\sim 4500$ Å. Hg e and especially Hg k already lie beyond the maximum of the absorption band. The length of the broken arrows qualitatively shows the lack of an exact proportionality between R intensity and absorption coefficient $\epsilon$. The diminution of quantum yield for R scattering apparently brought about by exaltation of the exciting frequency presumably has to be imputed to the simultaneous increase of radiationless processes starting from the intermediate level. The increase of the latter processes may secondarily lead to an augmentation of fluorescence yield (*cf.*, Pringsheim[6] p. 306*ff*).

### Simultaneous Appearance of Fluorescence

When 4-nitro-4'-dimethylaminostilbene (trans) in benzene solution is excited by Hg e, a broad fluorescence band appears in addition

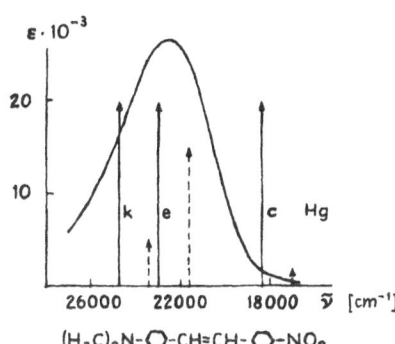

Fig. 19. Absorption spectrum of 4-nitro-4'-dimethylaminostilbene in pyridine solution after Shorygin and Ivanova.[72]

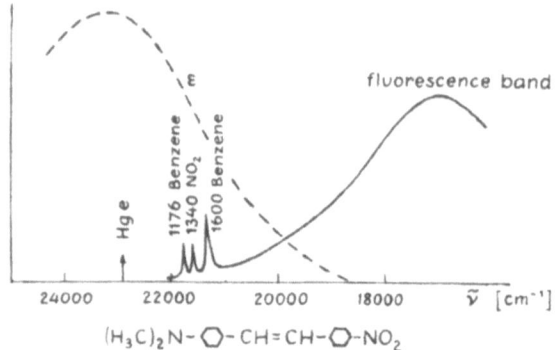

$(H_3C)_2N-\bigcirc-CH=CH-\bigcirc-NO_2$

Fig. 20. Simultaneous occurrence of RR scattering and fluorescence after Shorygin and Ivanova.[72,75]

to the R lines (Shorygin and Ivanova,[72,75] Fig. 20). Steps were taken by these authors to certify that the fluorescence emission starts from the same excited electronic level $e$ which is also acting as the intermediate level in the RE. At room temperature the bulk of fluorescence

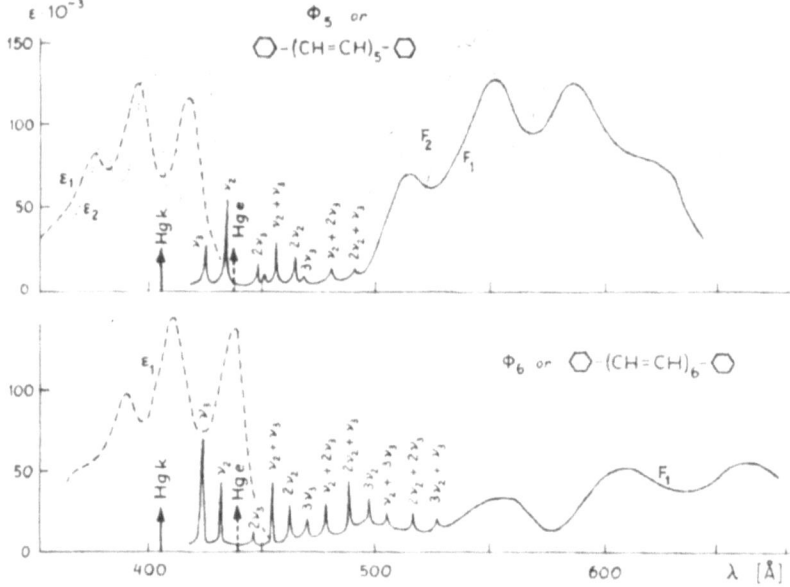

Fig. 21. RR and fluorescence spectra of diphenyldecapentene ($\phi_5$) and diphenyldodecahexene ($\phi_6$) after Shorygin and Ivanova.[76,77]. Acetone solutions at 25°C. Dotted lines: $\varepsilon_1$ and $\varepsilon_2$ are absorption spectra at 25 and $-70$°C, respectively; $F_2$ is the fluorescence spectrum at $-196$°C. R lines of solvent are omitted.

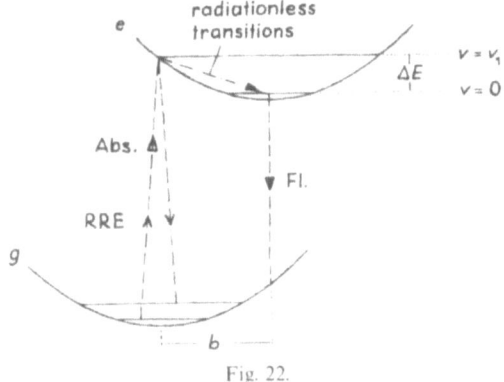

Fig. 22.

emission starts from the lowest vibrational level of $e$. Measurements of fluorescence afterglow showed that its lifetime is about $10^{-9}$ sec. On the other hand, the process of R scattering must elapse without any perceptible delay in the intermediate state $(e, v_1 \neq 0)$; the R lines, in contrast to the broadened structure of the fluorescence band, do not show any visible broadening.[75] Other examples of this phenomenon of simultaneous occurrence of scattering and fluorescence are presented by the Hg k-excited spectra of diphenyldecapentene (abbreviated $\phi_5$) and diphenyldodecahexene ($\phi_6$) in acetone solutions (Shorygin and Ivanova,[76,77] Fig. 21). These substances have absorption and fluorescence bands lying at rather different frequencies. This reveals a large separation $b$ of potential minima in states $g$ and $e$ (Fig. 22) and also a large difference $\Delta E$ of energy levels $c, 0$ and $e, v_1$. In substances where this is not the case, e.g., in naphthacene and pentacene, the RRE could not be observed.[77] This again seems to emphasize the fundamental importance of radiationless processes for the observability of the RRE.

## Characteristic Features

**Predictions of Theory.** The characteristic features of the rigorous RRE appear when the widths $2\delta_e^v$ of the vibration sublevels in the excited electronic state $e$ are smaller than their spacing (which apart from anharmonicity effects is identical to the corresponding vibrational frequency $\bar{v}$ in state $e$). In order that this condition be fulfilled, at least for some vibrational frequencies, the vibrational structure of the absorption band must be well visible.

The general theory being rather complicated, we will confine ourselves to a short discussion of the chief theoretical aspects. In the case that $\delta_e^v$ is negligibly small in comparison with $\bar{v}$ and that all assumptions basic to equation (9) can be made,* this equation for rigorous resonance excitation may be reduced to the form

$$\alpha_{gg}^{v_1 v_2} = \frac{1}{h} \frac{\mu_{eg}^{v'v_2}\mu_{ge}^{v_1 v'}}{v_{eg}^{v'v_1} - v_0 + i\delta_e^{v'}} \tag{33}$$

where $v'$ corresponds to that vibrational sublevel of $e$ which lies closest to the virtual level reached by irradiation of $v_0$ into the initial state $(g, v_1)$. In the zeroth approximation, the numerator of equation (33) is $\mu_{eg}(0)\mu_{ge}(0)(ev'|gv_2)(gv_1|ev')$. The last two factors in this expression are overlap integrals the values of which depend on the special form and relative situation of the two potential curves (surfaces) $g$ and $e$ and, therefore, cannot be given explicitly. It may easily be perceived, however, that in favorable circumstances the product of these two factors may assume considerable values even for large differences of quantum numbers $v_2$ and $v_1$. This means that in this case, contrary to the preresonance RE, even high overtones are likely to show up in the RR spectrum with intensities comparable to R fundamentals and to the Rayleigh line. A more extensive treatment shows that the same should hold for combination tones. The denominator of equation (33) indicates that the maximum intensity for all these lines is to be expected when $v_0 = v_{eg}^{v'v_1}$ and that its value depends on $\delta_e^{v'}$. The smaller $\delta_e^{v'}$, the larger the scattered intensities will be, but the quicker they will decrease with increase of $v_0 - v_{eg}^{v'v_1}$ (until finally for $v_0 - v_{eg}^{v'v_1} = \bar{v} = v_{eg}^{v''v_1} - v_{eg}^{v'v_1}$ there will again be resonance with the next sublevel $v''$). When $\delta_e^v$ is no longer negligible relative to $\bar{v}$, in equation (33) several terms from the sum over $v$ in equation (9) have to be retained, the superposition of which will result in a restriction (narrowing) of the overtone and combination tone structure of the RR spectrum.

Figure 23 after Shorygin and Krushinsky[78,79] qualitatively represents the general picture of resonance scattering spectra in the two cases $2\delta_e^v \gtrless v$ (only one vibrational frequency $v = \bar{v}$ identical for both states $g$ and $e$ presupposed).

---

* In addition, $2\delta_e^v$ must be presupposed to be larger than the frequency width of the exciting line. In the contrary case, which for customary excitation methods for R spectra would only occur occasionally with gases, the RE changes into what usually is called resonance fluorescence.

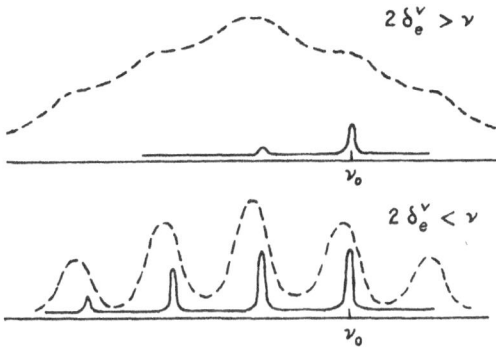

Fig. 23. General appearance of absorption curves (dotted lines) and resonance scattering spectra (solid lines) after Shorygin and Krushinsky.[78,79] $\nu_0$ is the exciting frequency (corresponding to the Rayleigh line).

### Experimental Observations

*Overtones and Combination Tones.* The R spectra reproduced in Fig. 21 show many overtones and combination tones. Table XV gives wave numbers and intensities for overtone and combination tone R lines of diphenyldodecahexene ($\phi_6$) in acetone solution excited by Hg e and Hg k, after Shorygin and Ivanova.[76,77] (Fundamentals

### TABLE XV
### Overtone and Combination Tone Raman Lines of Diphenyldodecahexene ($\phi_6$) After Shorygin and Ivanova[76,77]

| Raman wave number (cm$^{-1}$) | | Assignment | Raman band intensity (estimated) (IU) | |
|---|---|---|---|---|
| Hg e | Hg k | | Hg e | Hg k |
| 2282 | 2290 | $2\tilde{\nu}_3$ | 50 | * |
| 2690 | 2690 | $\tilde{\nu}_2 + \tilde{\nu}_3$ | 100 | 100 |
| 3100 | 3102 | $2\tilde{\nu}_2$ | * | * |
| 3425 | 3423 | $3\tilde{\nu}_3$ | 20 | 10 |
| 3821 | 3822 | $\tilde{\nu}_2 + 2\tilde{\nu}_3$ | 30 | 60 |
| 4232 | 4228 | $2\tilde{\nu}_2 + \tilde{\nu}_3$ | 30 | 100 |
| — | 4635 | $3\tilde{\nu}_2$ | — | 30 |
| — | 4951 | $\tilde{\nu}_2 + 3\tilde{\nu}_3$ | — | 20 |
| — | 5359 | $2\tilde{\nu}_2 + 2\tilde{\nu}_3$ | — | 30 |
| — | 5769 | $3\tilde{\nu}_2 + \tilde{\nu}_3$ | — | 20 |

* No intensity value given because of superposition of lines.

$\tilde{\nu}_3 \approx 1140\,\mathrm{cm}^{-1}$ and $\tilde{\nu}_2 \approx 1550\,\mathrm{cm}^{-1}$ refer to symmetric polyene-chain vibrations.) According to Fig. 21, Hg e still lies on the border (first vibrational maximum) of the absorption band; Hg k, however, lies in its interior. Therefore, the observable overtone structure by

## TABLE XVI
### Raman Spectra of $C_2H_5OOC(CH{=}CH)_nCOOC_2H_5$ (Abbreviated $X_n$), $n = 6, 7,$ and $8$, After Ivanova, Yanovskaya, and Shorygin[39]

| Hg e excitation | | Hg k excitation | | Assignment | $\Delta\tilde{\nu}$ (cm$^{-1}$) |
|---|---|---|---|---|---|
| Wave number (cm$^{-1}$) | Intensity $\times 10^6$(IU) | Wave number (cm$^{-1}$) | Intensity $\times 10^6$(IU) | | |
| **$X_6$** | | | | | |
| 1142 | 7 | 1142 | 105 | $\tilde{\nu}_3$ | — |
| 1562 | 17 | 1562 | 240 | $\tilde{\nu}_2$ | — |
| — | — | 2284, 2313 | 70 | $2\tilde{\nu}_3$ | 0 |
| — | — | 2701 | 70 | $\tilde{\nu}_2 + \tilde{\nu}_3$ | 2 |
| — | — | 3116 | — | $2\tilde{\nu}_2$ | 6 |
| — | — | 3425 | 10 | $3\tilde{\nu}_3$ | 1 |
| — | — | 3831 | 20 | $\tilde{\nu}_2 + 2\tilde{\nu}_3$ | 14 |
| — | — | 4253 | 20 | $2\tilde{\nu}_2 + \tilde{\nu}_3$ | 11 |
| **$X_7$** | | | | | |
| 1140 | 105 | 1140 | 55 | $\tilde{\nu}_3$ | — |
| 1550 | 237 | 1550 | 200 | $\tilde{\nu}_2$ | — |
| 2280 | 20 | 2280 | 13 | $2\tilde{\nu}_3$ | 0 |
| 2687 | 60 | 2687 | 80 | $\tilde{\nu}_2 + \tilde{\nu}_3$ | 3 |
| 3096 | 20 | 3096 | — | $2\tilde{\nu}_2$ | 4 |
| 3422 | 4 | 3422 | 13 | $3\tilde{\nu}_3$ | 2 |
| 3819 | 8 | 3817 | 43 | $\tilde{\nu}_2 + 2\tilde{\nu}_3$ | 12 |
| 4225 | 8 | 4222 | 65 | $2\tilde{\nu}_2 + \tilde{\nu}_3$ | 17 |
| — | — | 4631 | 22 | $3\tilde{\nu}_2$ | 19 |
| — | — | 4965 | 6 | $\tilde{\nu}_2 + 3\tilde{\nu}_3$ | 5 |
| — | — | 5355 | 13 | $2\tilde{\nu}_2 + 2\tilde{\nu}_3$ | 25 |
| — | — | 5763 | 8 | $3\tilde{\nu}_2 + \tilde{\nu}_3$ | 27 |
| **$X_8$** | | | | | |
| 1137 | 95 | 1139 | 72 | $\tilde{\nu}_3$ | — |
| 1540 | 260 | 1542 | 200 | $\tilde{\nu}_2$ | — |
| 2273 | 122 | — | — | $2\tilde{\nu}_3$ | 1 |
| 2670 | 140 | — | — | $\tilde{\nu}_2 + \tilde{\nu}_3$ | 7 |
| 3073 | 40 | — | — | $2\tilde{\nu}_2$ | 7 |
| 3410 | 45 | — | — | $3\tilde{\nu}_3$ | 1 |
| 3799 | 60 | — | — | $\tilde{\nu}_2 + 2\tilde{\nu}_3$ | 15 |
| 4200 | 43 | — | — | $2\tilde{\nu}_2 + \tilde{\nu}_3$ | 17 |
| 4916 | 5 | — | — | $\tilde{\nu}_2 + 3\tilde{\nu}_3$ | 35 |

excitation with Hg k is more expanded and intense than that by excitation with Hg e.

An identical observation was made by Ivanova, Yanovskaya, and Shorygin[39] on the diethyl ethers of polyene-$\alpha,\omega$-di-carbon acids $C_2H_5OOC(CH{=}CH)_nCOOC_2H_5$ (abbreviated $X_n$). Figure 24 shows the absorption curves for $X_6$, $X_7$, and $X_8$, and Table XVI the R wave numbers and intensities for these substances in acetone solutions excited by Hg e and Hg k. The most intense R lines are the fundamentals, overtones, and combination tones of the symmetric polyene-chain vibrations $\tilde{v}_3 \approx 1150\ cm^{-1}$ and $\tilde{v}_2 \approx 1550\ cm^{-1}$ (see Fig. 11). The last column in Table XVI gives the differences $\Delta\tilde{v}$ (due to anharmonicity) between overtone or combination tone wave numbers and the corresponding sum of fundamental tone wave numbers. Comparison of Table XVI and Fig. 24 shows that evidently an extended overtone structure in the RR spectrum may be observed only when the exciting frequency falls into the interior of the distinctly visible vibrational structure of the absorption band. It is interesting to note that in $X_8$ excited by Hg e the overtone $2v_3$ has an intensity higher than that of the fundamental $v_3$.

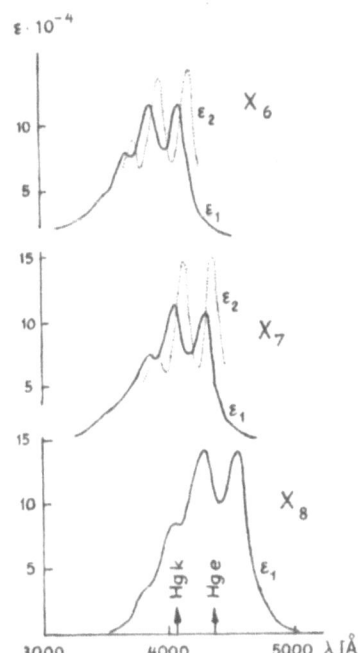

Fig. 24. First absorption bands of
$C_2H_5OOC(CH{=}CH)_nCOOC_2H_5$

(abbreviated $X_n$), $n = 6, 7$, and 8, in acetone solutions after Ivanova, Yanovskaya, and Shorygin.[39] Solid curve, $\varepsilon_1$ at 25°C; dotted curve, $\varepsilon_2$ at $-70$°C.

*Intensity Variations.* In the rigorous RRE the intensity of R lines (including overtone and combination tones) depends on the position of the exciting line relative to the single vibration maxima of the absorption band. Unfortunately, it is usually very difficult to resolve the total vibrational structure of the absorption band into the partial structures corresponding to the individual nuclear vibrations of a polyatomic molecule. Therefore, experimental evidence for the stated law may be given only in a few cases.

Table XVII after Shorygin and Ivanova[76,77] shows the intensity values $I_e$ excited by Hg e for the R lines $\tilde{v}_3$ and $\tilde{v}_2$ of compounds $\phi_5$ and $\phi_6$ in acetone solutions (see Fig. 21) and the intensity ratios $K = I_k/I_e$ for excitation by Hg k relative to that by Hg e. For $\phi_5$, $K > 1$; for $\phi_6$, $K < 1$. This is explained by the different ratios of absorption coefficient values $\epsilon_1$ (see Fig. 21) at the frequencies of Hg k and Hg e. Moreover, in both compounds $K$ differs for $\tilde{v}_3$ and $\tilde{v}_2$. This most probably can be attributed to the (unknown) difference in frequency positions of vibrational structure maxima for the individual vibrations $\tilde{v}_3$ and $\tilde{v}_2$.

The dotted lines marked $\epsilon_2$ and $F_2$ in Figs. 21 and 24 represent absorption and fluorescence curves at low temperatures. Obviously lowering of temperature leads to diminution of $\delta_e^v$ (i.e., prolongation of the lifetime of the vibrational states $v$ of $e$) and simultaneously to a shift of the absorption (fluorescence) bands towards longer (shorter) waves. These effects produce characteristic intensity variations of the individual RR lines, the shift of the absorption band being predominantly effective. This is verified by the data given in Table XVIII, which show the intensity variations of R lines $v_2$ and $v_3$ of compounds $X_6$ and $X_7$ in acetone solutions when the temperature is changed. Careful comparison of these very dissimilar intensity variations with

## TABLE XVII
### Values of Intensity Ratio for Compounds $\phi_5$ and $\phi_6$

| Compound | Wave number (cm$^{-1}$) | $I_e$ (IU) | $K = I_k/I_e$ |
|---|---|---|---|
| $\phi_5$ | $\tilde{v}_3 = 1145$ | $28 \times 10^6$ | 2.7 |
|  | $\tilde{v}_2 = 1567$ | $50 \times 10^6$ | 3.3 |
| $\phi_6$ | $\tilde{v}_3 = 1142$ | $35 \times 10^7$ | 0.65 |
|  | $\tilde{v}_2 = 1552$ | $53 \times 10^7$ | 0.25 |

## TABLE XVIII
Intensity Variation of Raman Bands $\tilde{v}_2 \approx 1550 \text{ cm}^{-1}$ and $\tilde{v}_3 \approx 1140$ $\text{cm}^{-1}$ in Acetone Solutions of $C_2H_5OOC(CH=CH)_nCOOC_2H_5$ (Abbreviated $X_n$), $n = 6$ and 7, by Lowering of Temperature After Ivanova, Yanovskaya, and Shorygin[39]

| Temperature (°C) | $X_6$ | | | | $X_7$ | | | |
|---|---|---|---|---|---|---|---|---|
| | Hg e | | Hg k | | Hg e | | Hg k | |
| | 1562* | 1142 | 1562 | 1142 | 1550 | 1140 | 1550 | 1140 |
| +35 | 1 | 1 | 1 | 1 | 1 | 1 | 1 | 1 |
| 0 | 1.1 | 1.4 | 0.8 | 0.9 | 1.1 | 1.2 | 0.7 | 1.3 |
| −40 | 1.8 | 2.0 | 0.6 | 0.7 | 1.3 | 1.5 | 0.5 | 2.0 |
| −95 | 3.1 | 4.0 | 0.3 | 0.6 | 1.0 | 1.3 | 0.4 | 2.4 |

* Values in this row are those of the Raman shift $(\text{cm}^{-1})$.

the shifts of vibrational structure maxima (Fig. 24) due to temperature change reveals that it is always the position of the exciting frequency relative to the single vibrational components of the absorption band which principally determines R intensities. This holds for fundamentals as well as for overtones and combination tones, and it must be considered a very satisfactory confirmation of theoretical predictions.

## CONCLUSION

We conclude this article with a short résumé of possible applications of the RRE. Of course, scientific application for investigation of material structure is the most important application, although under certain conditions the RRE might be of immediate use for routine work, such as mixture analysis. We note some special objects of research accessible to the RRE.

### Nature of Chemical Bond

Numerous publications have been dedicated to the investigation of bonding type by means of the RRE (see the numerous references (compiled up to 1956) in the work of Behringer and Brandmüller;[17] the work of Hester, Chapter 4 of this volume; and the work of numerous other authors.[18,34−36,40,50,52,54,73,74,84]) It is well known that quantum-mechanical resonance among molecular structures (conjugation, mesomeric effect) in connection with electronic charge interactions (inductive effect) not only influences bonding parameters, but also vibronic absorption spectra and, therefore, also must be reflected by the RRE.

## Molecular Structure

The RRE represents a supplementary method for investigation of molecular structure. The variation of R line characteristics (frequency, intensity, depolarization ratio) by change of molecular constitution is, of course, always accompanied by variations of other molecular parameters, such as dipole moment, refractivity, and absorbance; but, in the case of resonance, the examination of R intensities usually proves to be the most sensitive test for structural changes. It may be expected that in future work the RRE, in contradistinction to the ordinary RE, also will help to investigate excited molecular states, thus complementing the methods of ultraviolet and visible absorption and fluorescence spectroscopy.

## Intermolecular Interactions

The RRE reacts very sensitively even to small shifts and alterations of molecular electronic levels such as are caused by changes of inter-molecular forces due to solvent effects, temperature variations, or transitions to other states of matter.

## NOTATION

### Raman Terms

R    = Raman
RE   = Raman effect
RR   = Resonance Raman
RRE = Resonance Raman effect

### Hg Lines

$a = 5791$ Å or $17{,}264 \text{ cm}^{-1}$
$b = 5770$ Å or $17{,}327 \text{ cm}^{-1}$
$c = 5461$ Å or $18{,}307 \text{ cm}^{-1}$
$e = 4358$ Å or $22{,}938 \text{ cm}^{-1}$
$i = 4078$ Å or $24{,}516 \text{ cm}^{-1}$
$k = 4047$ Å or $24{,}705 \text{ cm}^{-1}$
$g = 3650$ Å or $27{,}388 \text{ cm}^{-1}$

# REFERENCES

1. J. Brandmüller and H. Moser, *Einführung in die Ramanspektroskopie*, Dr. Dietrich Steinkopff Verlag, Darmstadt, Germany, 1962.
2. G. Geppert, *Experimentelle Methoden der Molekülspektroskopie*, Akademie-Verlag, Berlin, Germany, 1964.
3. W. Heitler, *The Quantum Theory of Radiation*, Clarendon Press, Oxford, England, 1954.
4. G. Herzberg, *Molecular Spectra and Molecular Structure, Vol. 2: Infrared and Raman Spectra of Polyatomic Molecules*, D. Van Nostrand Company, Inc., Princeton, New Jersey, 1956.
5. K. W. F. Kohlrausch, *Ramanspektren*, Akademische Verlagsgesellschaft, Leipzig, Germany, 1943.
6. P. Pringsheim, *Fluorescence and Phosphorescence*, Interscience Publishers Inc., New York, 1949.
7. C. Sandorffy, *Die Elektronenspektren in der Theoretischen Chemie*, Verlag Chemie, G.m.b.H., Weinheim/Bergstrasse, Germany, 1961.
8. G. W. Wheland, *Resonance in Organic Chemistry*, John Wiley & Sons, Inc., New York, and Chapman & Hall Ltd., London, 1955.
9. E. B. Wilson, J. C. Decius, and P. C. Cross, *Molecular Vibrations*, McGraw-Hill Book Company, New York–Toronto–London, 1955.
10. J. Behringer, "Über die Absorptionskorrektur in Resonanz-Raman-Spektren," in: *Advances in Molecular Spectroscopy, Vol. 3*, Pergamon Press, Oxford, England, 1962, p. 1087.
11. G. Placzek, "Rayleigh-Streuung und Raman-Effekt," in: E. Marx (ed.), *Handbuch der Radiologie, Vol. 6*, Akademische Verlagsgesellschaft, Leipzig, Germany, 1934, Part 2, p. 209.
12. G. Wentzel, "Wellenmechanik der Stoss- und Strahlungsvorgänge," in: H. Geiger and K. Scheel (eds.), *Handbuch der Physik, Vol. 24*, Springer-Verlag, Berlin, 1933, Part 1, p. 695.
13. F. Vratny and R. B. Fischer, The Effect of Absorbancy upon Raman Intensities in Solutions," in: *Talanta, Vol. 3*, Pergamon Press, Oxford, England, 1959, p. 315.
14. A. C. Albrecht, *J. Chem. Phys.* **33**: 156 (1960).
15. A. C. Albrecht, *J. Chem. Phys.* **34**: 1476 (1961).
16. P. A. Bazhulin and M. M. Sushchinsky, *Usp. Fiz. Nauk* **68**: 135 (1959)
17. J. Behringer and J. Brandmüller, *Z. Elektrochem.* **60**: 643 (1956).
18. J. Behringer, thesis, University of Munich (1957).
19. J. Behringer, *Z. Elektrochem.* **62**: 544 (1958).
20. J. Behringer, *Z. Elektrochem.* **62**: 906 (1958).
21. J. Behringer and J. Brandmüller, *Ann. Physik* **4** (7): 234 (1959).
22. J. Behringer and J. Brandmüller, *Z. Angew. Phys.* **14**: 674 (1962).
23. Ya. S. Bobovich and V. V. Perekalin, *Dokl. Akad. Nauk SSSR* **124**: 1239 (1959).
24. Ya. S. Bobovich and M. Ya. Tsenter, *Opt. i Spektroskopiya* **8**: 45 (1960).
25. Ya. S. Bobovich and Ya. A. Eidus, *Opt. i Spektroskopiya* **16**: 424 (1964).
26. J. Brandmüller, *Optik* **12**: 389 (1955).
27. J. Brandmüller and H. Moser, *Z. Angew. Phys.* **9**: 567 (1957).
28. J. Brandmüller, E. W. Schmid, and H. W. Schrötter, *Spectrochim. Acta* **17**: 523 (1961).
29. R. Eckert and H. Kuhn, *Z. Elektrochem.* **64**: 356 (1960).
30. A. I. Finkelstein, thesis, Moscow (1950).
31. A. I. Finkelstein and P. P. Shorygin, *Zh. Fiz. Khim.* **26**: 1272 (1952).
32. D. Fisher and E. R. Lippincott, *Spectrochim. Acta* **6**: 255 (1954).
33. R. D. Fisher, M. S. thesis, Kansas State University, Manhattan, Kansas (1954).
34. M. Harrand, *Ann. Phys. (Paris)* **8** (12): 126 (1953).

35. M. Harrand and H. Martin, *Bull. Soc. Chim. France* 1383 (1956).
36. M. Harrand and C. Bazin, *J. Phys. Radium* **18**: 687 (1957).
37. W. Hofmann and H. Moser, *Z. Elektrochem.* **64**: 310 (1960).
38. W. Hofmann and H. Moser, *Ber. Bunsenges. Physik. Chem.* **68**: 129 (1964).
39. T. M. Ivanova, L. A. Yanovskaya, and P. P. Shorygin, *Opt. i Spektroskopiya* **18**: 206 (1965).
40. F. Jost and M. Harrand, *J. Phys. Radium* **23**: 308 (1962).
41. Z. Kecki, *Prace Konf. Elektrochem. Warsaw*, 687 (1957).
42. A. Kh. Khalilov, thesis, V.I. Lenin Teaching Institute, Moscow (1951).
43. A. Kh. Khalilov, *Izv. Akad. Nauk SSSR, Ser. Fiz.* **17**: 586 (1953).
44. H. B. Klevens and J. R. Platt, *J. Chem. Phys.* **17**: 470 (1949).
45. I. I. Kondilenko, P. A. Korotkov, and V. L. Strizhevsky, *Opt. i Spektroskopiya* **9**: 26 (1960).
46. I. I. Kondilenko and V. L. Strizhevsky, *Opt. i Spektroskopiya* **11**: 262 (1961).
47. I. I. Kondilenko and V. L. Strizhevsky, *Opt. i Spektroskopiya* **18**: 938 (1965).
48. R. Lippincott, J. P. Sibilia, and R. D. Fisher, *J. Opt. Soc. Am.* **49**: 83 (1959).
49. W. Maier and F. Dörr, *Appl. Spectr.* **14**: 1 (1960).
50. W. Maier, thesis, Technische Hochschule, Munich (1959).
51. N. T. Melamed, *J. Appl. Phys.* **34**: 560 (1963).
52. G. Michel and G. Duyckaerts, *Spectrochim. Acta* **10**: 259 (1958).
53. G. Michel, *Spectrochim. Acta* **12**: 400 (1958).
54. G. Michel, *Bull. Soc. Chim. Belges* **68**: 643 (1959).
55. H. Moser and J. Varchmin, *Z. Angew. Phys.* **17**: 392 (1964).
56. E. M. Popov and G. A. Kogan, *Opt. i Spektroskopiya* **17**: 670 (1964).
57. D. G. Rea, *J. Opt. Soc. Am.* **49**: 90 (1959).
58. D. G. Rea, *J. Mol. Spectry.* **4**: 999 (1960).
59. F. A. Savin and I. I. Sobelman, *Opt. i Spektroskopiya* **7**: 733 (1959).
60. F. A. Savin and I. I. Sobelman, *Opt. i Spektroskopiya* **7**: 740 (1959).
61. F. A. Savin, *Opt. i Spektroskopiya* **15**: 42 (1963).
62. P. P. Shorygin, *Zh. Fiz. Khim.* **15**: 1075 (1941).
63. P. P. Shorygin and M. V. Volkenstein, *Izv. Akad. Nauk SSSR, Ser. Fiz.* **5**: 174 (1941).
64. P. P. Shorygin, *Usp. Khim.* **19**: 419 (1950).
65. P. P. Shorygin, *Dokl. Akad. Nauk SSSR* **78**: 469 (1951).
66. P. P. Shorygin, *Dokl. Akad. Nauk SSSR* **87**: 201 (1952).
67. P. P. Shorygin, *Izv. Akad. Nauk SSSR, Ser. Fiz.* **17**: 581 (1953).
68. P. P. Shorygin, *J. Chim. Phys.* **50**: D31 (1953).
69. P. P. Shorygin and L. Z. Osityanskaya, *Dokl. Akad. Nauk SSSR* **98**: 51 (1954).
70. P. P. Shorygin and L. Z. Osityanskaya, *Izv. Akad. Nauk SSSR, Ser. Fiz.* **18**: 681 (1954).
71. P. P. Shorygin and L. Krushinsky, *Z. Physik* **150**: 332 (1958).
72. P. P. Shorygin and T. M. Ivanova, *Dokl. Akad. Nauk SSSR* **121**: 70 (1958).
73. P. P. Shorygin and Z. S. Yegorova, *Dokl. Akad. Nauk SSSR* **118**: 763 (1958).
74. P. P. Shorygin, T. N. Shkurina, M. C. Shostakovsky, F. P. Sidelkovskaya, and M. G. Zelenskaya, *Izv. Akad. Nauk SSSR, Otd. Khim. Nauk* 2208 (1959).
75. P. P. Shorygin, *Pure Appl. Chem.* **4**: 87 (1962).
76. P. P. Shorygin and T. M. Ivanova, *Dokl. Akad. Nauk SSSR* **150**: 533 (1963).
77. P. P. Shorygin and T. M. Ivanova, *Opt. i Spektroskopiya* **15**: 176 (1963).
78. P. P. Shorygin and L. L. Krushinsky, *Opt. i Spektroskopiya* **17**: 551 (1964).
79. P. P. Shorygin and L. L. Krushinsky, *Dokl. Akad. Nauk SSSR* **154**: 571 (1964).
80. H. W. Schrötter, *Z. Elektrochem.* **64**: 853 (1960).
81. H. W. Schrötter, *Z. Angew. Phys.* **12**: 275 (1960).
82. H. Sponer and E. Teller, *Rev. Mod. Phys.* **13**: 76 (1941).
83. M. M. Sushchinsky and V. A. Zubov, *Opt. i Spektroskopiya* **13**: 766 (1962).

84. B. Théry, M. Harrand, and P. Grammaticakis, *J. Phys. Radium* **24**: 297 (1963).
85. M. Ya. Tsenter and Ya. S. Bobovich, *Opt. i Spektroskopiya* **12**: 54 (1962).
86. M. Ya. Tsenter and Ya. S. Bobovich, *Opt. i Spektroskopiya* **16**: 246 (1964).
87. M. Ya. Tsenter and Ya. S. Bobovich, *Opt. i Spektroskopiya* **16**: 417 (1964).
88. F. Vratny and R. B. Fischer, *Appl. Spectry.* **14**: 76 (1960).
89. F. Vratny, *Anal. Chim. Acta* **23**: 171 (1960).
90. H. Wittek, *Monatsh. Chem.* **73**: 231 (1941).
91. V. A. Zubov, *Opt. i Spektroskopiya* **13**: 861 (1962).

## Chapter 7

# Raman Spectroscopy of Complex Ions in Solution

## D. E. Irish

*Department of Chemistry*
*University of Waterloo*
*Waterloo, Ontario, Canada*

## INTRODUCTION

A complete understanding of complex ions in solution depends on knowledge of (1) the identity of the complexes; (2) the structures of the species; (3) the equilibrium constants; (4) the thermodynamics of formation; and (5) the mechanisms of formation of the species. Photoelectric Raman spectroscopy is a powerful technique for obtaining some of this information. The number of Raman bands and their degrees of depolarization are often sufficient to establish both the identity and the structure of the species. In principle, infrared data are necessary for a complete vibrational analysis, but for many of the simple structures involved, an unambiguous conclusion can be drawn without infrared data. Many of the vibrational modes of metal complexes are at frequencies less than $400 \, \text{cm}^{-1}$. Raman spectrophotometers provide useful information even down to $40 \, \text{cm}^{-1}$, but specialized equipment is needed for infrared studies in this spectral region. Furthermore, for many studies of interest to date, the complex ions exist in aqueous solution. These systems present no special difficulty for the Raman technique. On the other hand, infrared studies, especially quantitative studies, cannot easily be made on these systems, although the development of new cell-window materials and the attenuated total reflectance technique will probably make more infrared data available in the future. If the mercury 4358-Å line is used for Raman excitation, investigation is restricted to colorless or nearly colorless systems. The development of sources which utilize other excitation frequencies is enlarging the number and types of systems which can be investigated.

The relative integrated intensities of Raman bands can be empirically related to the concentration of Raman-active species; these intensities are often, within narrow limits, directly proportional to the ratio of the concentrations. This quantitative aspect of Raman spectroscopy can provide data for the evaluation of equilibrium constants, thermodynamic functions, and reaction rates.

In Part 1 of this chapter, the selection rules for the common structures will be summarized and the methods used for evaluating equilibrium constants, thermodynamic properties, and reaction rates will be discussed. Part 2 will be devoted to a review of the frequencies of many of the simple complex ions studied to date. The examples will in general be restricted to complexes found in a solution where the central group is a metal ion M, which is associated with a ligand A to form a complex of the general form $M_mA_n$. It will also be desirable to include as examples some representative anionic complexes of the general form $M_mA_n$ where M designates a central group which is a nonmetal. Although the chapter will in general be restricted to Raman studies, infrared data will be included whenever necessary for the argument.

# PART 1

## IDENTITY AND GEOMETRY

### The Selection Rules for Common Point Groups

The internal motions of a vibrating molecule, although apparently complex, can be resolved into a number of relatively simple vibratory motions known as the normal modes of vibration. Each of these modes has a characteristic frequency dependent on the masses of atoms and the forces between them. The number of modes is $3N - 6$ for a nonlinear molecule and $3N - 5$ for a linear molecule containing $N$ atoms. The activity of the normal modes in Raman or infrared spectroscopy is dependent on the geometry of the molecule. The shape of a molecule is specified by the symmetry elements that can be ascribed to it (axes of rotation, planes of symmetry, center of symmetry), which leads to a point-group notation. The reader is referred to standard works for the method of classifying molecules by point groups.[1,2]

## TABLE I
### Number and Species of Fundamental Modes of Vibration for Common Point Groups

| Molecular type | Point group | Description | Reduced representation | $n_R$ | $n_P$ | $n_I$ | $n_C$ |
|---|---|---|---|---|---|---|---|
| 1. MA | $C_{\infty v}$ | Linear | $\Sigma^+(R, I)$ | 1 | 1 | 1 | 1 |
| 2. MA$_2$ | $D_{\infty h}$ | Linear | $\Sigma_g^+(R) + \Sigma_u^+(I) + \Pi_u(I)$ | 1 | 1 | 2 | 0 |
| 3. MAB | $C_{2v}$ | Bent | $2A_1(R, I) + B_1(R, I)$ | 3 | 2 | 3 | 3 |
| | $C_{\infty v}$ | Linear | $2\Sigma^+(R, I) + \Pi(R, I)$ | 3 | 2 | 3 | 3 |
| | $C_s$ | Bent | $2A'(R, I) + A''(R, I)$ | 3 | 3 | 3 | 3 |
| 4. MA$_3$ | $D_{3h}$ | Centered equilateral triangle | $1A_1'(R) + 1A_2''(I) + 2E'(R, I)$ | 3 | 1 | 3 | 2 |
| 5. BMA$_2$ | $C_{3v}$ | Pyramidal | $2A_1(R, I) + 2E(R, I)$ | 4 | 2 | 4 | 4 |
| | $C_{2v}$ | Planar Y | $3A_1(R, I) + 2B_1(R, I) + B_2(R, I)$ | 6 | 3 | 6 | 6 |
| | $C_s$ | Pyramidal | $4A'(R, I) + 2A''(R, I)$ | 6 | 4 | 6 | 6 |
| 6. AM$_2$A | $D_{\infty h}$ | Linear | $2\Sigma_g^+(R) + \Sigma_u^+(I) + \Pi_g(R) + \Pi_u(I)$ | 3 | 2 | 2 | 0 |
| | $C_{2h}$ | Planar cis form | $3A_1(R, I) + A_2(R) + 2B_1(R, I)$ | 6 | 3 | 5 | 5 |
| | $C_{2h}$ | Planar trans form | $3A_g(R) + A_u(I) + 2B_u(I)$ | 3 | 3 | 3 | 0 |
| | $C_{2v}$ | Planar Y | $3A_1(R, I) + 2B_1(R, I) + B_2(R, I)$ | 6 | 3 | 6 | 6 |
| M$_2$A$_2$ | $C_2$ | Nonplanar twisted | $4A(R, I) + 2B(R, I)$ | 6 | 4 | 6 | 6 |
| 7. MA$_4$ | $T_d$ | Tetrahedral | $A_1(R) + E(R) + 2F_2(R, I)$ | 4 | 1 | 2 | 2 |
| | $D_{4h}$ | Square planar | $A_{1g}(R) + A_{2u}(I) + B_{1g}(R) + B_{2g}(R) + B_{2u}(ia) + 2E_u(I)$ | 3 | 1 | 3 | 0 |
| 8. M(AB)$_4$ | $C_{4v}$ | Tetragonal pyramid | $2A_1(R, I) + 2B_1(R) + B_2(R) + 2E(R, I)$ | 7 | 2 | 4 | 4 |
| | $T_d$ | Tetrahedral | $2A_1(R) + 2E(R) + F_1(ia) + 4F_2(R, I)$ | 8 | 2 | 4 | 4 |
| | $D_{4h}$ | Square planar | $2A_{1g}(R) + A_{2g}(ia) + 2B_{1g}(R) + 2B_{2g}(R) + E_g(R) + 2A_{2u}(I) + B_{1u}(ia) + B_{2u}(ia) + 4E_u(I)$ | 7 | 2 | 6 | 0 |
| 9. MA$_3$B | $C_{3v}$ | Tetrahedral | $3A_1(R, I) + 3E(R, I)$ | 6 | 3 | 6 | 6 |
| | $C_{2v}$ | Planar with $C_2$ axis | $4A_1(R, I) + 3B_1(R, I) + 2B_2(R, I)$ | 9 | 4 | 9 | 9 |
| | $C_s$ | Planar | $7A'(R, I) + 2A''(R, I)$ | 9 | 7 | 9 | 9 |
| 10. MA$_2$B$_2$ | $C_{2v}$ | Tetrahedral | $4A_1(R, I) + A_2(R) + 2B_1(R, I) + 2B_2(R, I)$ | 9 | 4 | 8 | 8 |
| | $D_{2h}(V_h)$ | trans Symmetric planar | $2A_g(R) + B_{1g}(R) + 2B_{1u}(I) + 2B_{2u}(I) + 2B_{3u}(I)$ | 3 | 2 | 6 | 0 |
| | $C_{2v}$ | cis Symmetric planar | $4A_1(R, I) + A_2(R) + 3B_1(R, I) + B_2(R, I)$ | 9 | 4 | 8 | 8 |

| | | | | | | | |
|---|---|---|---|---|---|---|---|
| 11. $MA_2(AB)$ | $C_s$ | Planar | $7A'(R,I) + 2A''(R,I)$ | 9 | 7 | 9 | 9 |
| 12. $MA_5$ | $D_{3h}$ | Trigonal bipyramid | $2A_1'(R) + 2A_2''(I) + 3E'(R,I) + E''(R)$ | 6 | 2 | 5 | 3 |
| 13. $MA_4B$ | $C_{4v}$ | Tetragonal pyramid | $3A_1(R,I) + 2B_1(R) + B_2(R) + 3E(R,I)$ | 9 | 3 | 6 | 6 |
| | $C_{3v}$ | Trigonal bipyramid; B axial | $4A_1(R,I) + 4E(R,I)$ | 8 | 4 | 8 | 8 |
| | $C_{2v}$ | Trigonal bipyramid; B equatorial | $5A_1(R,I) + A_2(R) + 3B_1(R,I) + 3B_2(R,I)$ | 12 | 5 | 11 | 11 |
| 14. $MA_3B_2$ | $C_{4v}$ | Tetragonal pyramid | $3A_1(R,I) + 2B_1(R) + B_2(R) + 3E(R,I)$ | 9 | 3 | 6 | 6 |
| | $D_{3h}$ | Trigonal bipyramid; B axial | $2A_1'(R) + 2A_2''(I) + 3E'(R,I) + E''(R)$ | 6 | 2 | 5 | 3 |
| | $C_s$ | Distorted trigonal bipyramid | $8A'(R,I) + 4A''(R,I)$ | 12 | 8 | 12 | 12 |
| | $C_{2v}$ | Trigonal bipyramid; B equatorial | $5A_1(R,I) + A_2(R) + 3B_1(R,I) + 3B_2(R,I)$ | 12 | 5 | 11 | 11 |
| 15. $A_2M_2A_2$ | $D_{2h}(V_h)$ | Symmetric planar | $3A_g(R) + A_u(ia) + 2B_{1g}(R) + B_{1u}(I) + B_{2g}(R) + 2B_{2u}(I) + 2B_{3u}(I)$ | 6 | 3 | 5 | 0 |
| 16. $ABM_2AB$ | $C_{2v}$ | cis Symmetric planar | $5A_1(R,I) + 2A_2(R) + 4B_1(R,I) + B_2(R,I)$ | 12 | 5 | 10 | 10 |
| | $C_{2h}$ | trans Symmetric planar | $5A_g(R) + 2A_u(I) + B_g(R) + 4B_u(I)$ | 6 | 5 | 6 | 0 |
| 17. $A_2M_2B_2$ | $C_{2v}$ | Unsymmetric planar | $5A_1(R,I) + A_2(R) + 4B_1(R,I) + 2B_2(R,I)$ | 12 | 5 | 11 | 11 |
| 18. $MA_6$ | $O_h$ | Regular octahedron | $A_{1g}(R) + E_g(R) + 2F_{1u}(I) + F_{2g}(R) + F_{2u}(ia)$ | 3 | 1 | 2 | 0 |
| | $D_{4h}$ | Distorted octahedron | $2A_{1g}(R) + 2A_{2u}(I) + B_{1g}(R) + B_{2g}(R) + B_{2u}(ia) + E_g(R) + 3E_u(I)$ | 5 | 2 | 5 | 0 |
| | $D_{6h}$ | Plane symmetric | $A_{1g}(R) + A_{2u}(I) + B_{1g}(ia) + B_{1u}(ia) + B_{2u}(ia) + 2E_{1u}(I) + 2E_{2g}(R) + E_{2u}(ia)$ | 3 | 1 | 3 | 0 |
| 19. $MA_5B$ | $C_{4v}$ | Octahedron | $4A_1(R,I) + 2B_1(R) + B_2(R) + 4E(R,I)$ | 11 | 4 | 8 | 8 |
| $MA_4B_2$ | $D_{4h}$ | trans Octahedron | $2A_{1g}(R) + 2A_{2u}(I) + B_{1g}(R) + B_{2g}(R) + B_{2u}(ia) + E_g(R) + 3E_u(I)$ | 5 | 2 | 5 | 0 |
| 20. $MA_3B_3$ | $C_{2v}$ | cis Octahedron | $6A_1(R,I) + 2A_2(R) + 4B_1(R,I) + 3B_2(R,I)$ | 15 | 6 | 13 | 13 |
| | $C_s$ | Octahedron; 2B's in plane | $9A'(R,I) + 6A''(R,I)$ | 15 | 9 | 15 | 15 |
| | $C_{2v}$ | Octahedron; 2B's in plane | $6A_1(R,I) + A_2(R) + 4B_1(R,I) + 4B_2(R,I)$ | 15 | 6 | 14 | 14 |
| 20. $M(AB)_6$ | $O_h$ | Regular octahedron | $2A_{1g}(R) + 2E_g(R) + F_{1g}(ia) + 4F_{1u}(I) + 2F_{2g}(R) + 2F_{2u}(ia)$ | 6 | 2 | 4 | 0 |
| 21. $A_2MA_2MA_2$ | $D_{2h}(V_h)$ | Bridged | $4A_g(R) + A_u(ia) + 2B_{1g}(R) + 3B_{1u}(I) + 2B_{2g}(R) + 2B_{2u}(I) + B_{3g}(R) + 3B_{3u}(I)$ | 9 | 4 | 8 | 0 |

The normal modes of vibration also possess symmetry properties related to the symmetry of the molecule and are classified according to these symmetries. Such a group-theoretical analysis does not require knowledge of the frequencies or the detailed forms of the normal modes. For an isolated molecule, the analysis reveals the number of normal modes and their symmetry type, the number of Raman active modes and infrared active modes, the number of polarized Raman lines, and the activity of overtones and combination bands. A vibration appears as a fundamental in the Raman spectrum if its symmetry is the same as that of one of the six components of the polarizability. The degree of depolarization of the band is less than $\frac{6}{7}$ if the vibration is totally symmetric. A vibration appears as a fundamental in the infrared spectrum if its symmetry is the same as that of one of the three components of the dipole moment. The application of group-theoretical methods to molecular vibrations is beyond the scope of this chapter and is well presented elsewhere.[1,3-5] The reduced representation of internal motion and related information is stated without proof in Table I for point groups commonly encountered in coordination chemistry (cf. Wilson[6]). The terms $R$, $I$, and $ia$ designate Raman-active, infrared-active, and inactive, respectively. The term $n_R$ denotes the number of Raman-active bands, and $n_P$ denotes the number of these which are polarized. The term $n_I$ denotes the number of infrared-active bands, and $n_C$ denotes the number of bands which have coincident frequencies in the Raman and infrared spectra. The total number of active fundamentals is then given by $(n_R + n_I - n_C)$.

As an illustration of the use of the table, let us consider the complex ion $[Pt(CN)_4]^{-2}$. This complex might be considered to have four cyanide ligands tetrahedrally arranged around platinum, in which case the point group would be $T_d$; or the four cyanide ligands could be located at the corners of a platinum-centered square (point group $D_{4h}$). Reference to Table I reveals that for $M(AB)_4$ the following would apply:

$$T_d \ldots 8 \text{ Raman lines, } \quad 2 \text{ polarized}$$

$$D_{4h} \ldots 7 \text{ Raman lines, } \quad 2 \text{ polarized}$$

On the basis of the number of observed Raman lines a decision could be made in favor of one of the chosen geometries. In this particular case, the spectral differences for the two geometries are not large and knowledge of the infrared spectrum would be desirable. The table also indicates that four infrared lines should be coincident with four

Raman lines if the symmetry is $T_d$, but that there should be no coincidences if the symmetry is $D_{4h}$. Information concerning the number of coincidences would therefore verify the prediction. The observed spectrum is in accord with $D_{4h}$ symmetry.[7]

When the molecule is not isolated, the environment can have a marked influence on the spectrum. The selection rules described above may not be obeyed in a condensed phase. Halford[8-10] has pointed out that the symmetry which is important for predicting the activity of the normal modes of crystals is the site symmetry; which is defined as the symmetry of the potential field in which the entity under observation vibrates. For a crystal, the site symmetry can be deduced from the space group and the number of molecules contained in the unit cell. Halford[8] has compiled the requisite information for 230 space groups. As one example of the application of this approach, the reader is referred to the spectral analysis of some hexafluorogermanates.[11]

Halford[8,9] suggests that no selection rules operate in the liquid phase. For solutions, the situation is dependent on the magnitude of the solute–solvent interactions. For moderately dilute aqueous solutions, the interaction between water molecules and most complex ions is weak and the spectra of the complex ions are generally in accord with the selection rules for isolated species. For electrolyte solutions, ions can interact without significant covalent-bond formation. An interaction between oppositely charged ions through intervening solvent molecules of the first coordination sphere of one or both ions is referred to as an ion pair.[12] For a polyatomic ion in this situation, the symmetry of the potential field resulting from the environment may be lower than the point-group symmetry and degenerate vibrations may be split into resolvable bands. Thus, the spectrum of the nitrate ion in aqueous calcium nitrate solution suggests that this ion is in a potential field of $C_{2v}$ symmetry due to an adjacent calcium ion[13] (possibly a solvated calcium ion[14]), not $D_{3h}$ symmetry typical of an isolated nitrate ion. For ion pairs, the metal–anion vibration is not usually detected. George, Rolfe, and Woodward[15] concluded that the intensity to be expected from an ion pair is $10^{-2}$ to $10^{-5}$ times as large as the intensity from covalently linked atoms. For nonaqueous solutions, ion–solvent interactions are often detected and may be more prominent spectroscopically than ion–ion interactions. Thus, although ion-pair formation is detectable in acetonitrile solutions of silver nitrate, a silver-acetonitrile complex is predominant.[16]

Thus, a complex ion in solution can be detected and identified by correlation of Raman and infrared spectral data with the selection rules for probable point groups. The analysis often yields a structure for the species. However, the investigator must also consider the lowering of symmetry because of ionic interactions, solute–solvent effects, and the failure of selection rules in condensed phase.

### Identification from Intensity Studies

The variation of the intensity of a Raman line characteristic of a particular species with the composition of a solution can sometimes be employed to establish the identity of the species (complex ion or ion pair) which generates the line. The Job method of continuous variations, attributed to Job[17] and extended by Vosburgh and Cooper,[18] will be described.

Consider a single complex ion formed by the reaction

$$M + nA \rightleftharpoons MA_n$$

A series of solutions is prepared in which the sum of the formal concentrations of M and A is constant but the ratio of the concentrations varies. Then,

$$C_M + C_A = C$$

Define the fraction $f$ such that in any one of the solutions

$$f = C_A/C$$

Then,

$$C_M = (1 - f)C$$

In any one of the solutions we can calculate the concentrations of the uncombined M or A:

$$[M] = (1 - f)C - [MA_n]$$

$$[A] = fC - n[MA_n]$$

The concentration of the complex will change as $f$ is varied. It will be zero when $f$ is either zero or one, and it will have a maximum at some intermediate value of $f$. If the intensity of a Raman line characteristic of only the complex is measured and plotted *versus* $f$, the maximum intensity will occur at the value of $f$ which corresponds to the maximum concentration of $MA_n$. This value of $f$ can be used to determine $n$ as follows:

This value of $f$ occurs at the point where $d[MA_n]/df$ is zero

$$Q_c = \frac{[MA_n]}{[M][A]^n}$$

where $Q_c$ is the concentration quotient which occurs in the expression of the equilibrium constant. Hence,

$$[MA_n] = Q_c\{(1 - f)C - [MA_n]\}\{fC - n[MA_n]\}^n \qquad (1)$$

If the assumption is made that $Q_c$ is essentially constant for all solutions of the series, and if the first derivative with respect to $f$ is set equal to zero, the relation simplifies to

$$n = \frac{f}{1 - f}$$

Thus, if the maximum intensity occurs at $f = 0.5$, a $1:1$ complex is inferred; if the maximum intensity occurs at $f = 0.67$, the complex is of the type $MA_2$, and so on. The Raman intensity could also be plotted against $C_A/C_M$. The stoichiometry of the complex is then the value of $C_A/C_M$ which corresponds to the maximum. This plot is advantageous if $n$ is 4, 5, or 6 because the probable positions of the maxima are evenly spaced, whereas on the intensity *versus* $f$ plot the positions for the maxima occur at points that become increasingly closer together, i.e., for $f$ values of 0.80, 0.833, and 0.857.

The shape of the curve and the sharpness of the maximum depend on the stability of the complex. A broad flat maximum and considerable curvature are indicative of a small association constant. If the complex is only slightly dissociated, the graph will consist of two almost straight lines and a sharp maximum. The requirement that $Q_c$ be a constant for the series of solutions is perhaps the weakness of the method, particularly if used in conjunction with Raman line intensities when concentrated solutions are involved. However, Nixon and Plane[19] have reported the relative constancy of $Q_c$ with changing solution composition for a number of concentrated electrolyte systems.

One example of the application of the Job method is the determination of the origin of the 740 cm$^{-1}$ band in aqueous calcium nitrate solutions.[14] The band intensity is at a maximum when the $[Ca^{+2}]/[NO_3^-]$ ratio is 1 (Fig. 1). The band has been interpreted as characteristic of an ion pair $(Ca-H_2O-NO_3)^+$. The perturbation of the nitrate ion by the hydrated calcium ion results in a lowering of the symmetry from $D_{3h}$ to $C_{2v}$ and the removal of the degeneracy of the $\nu_4$

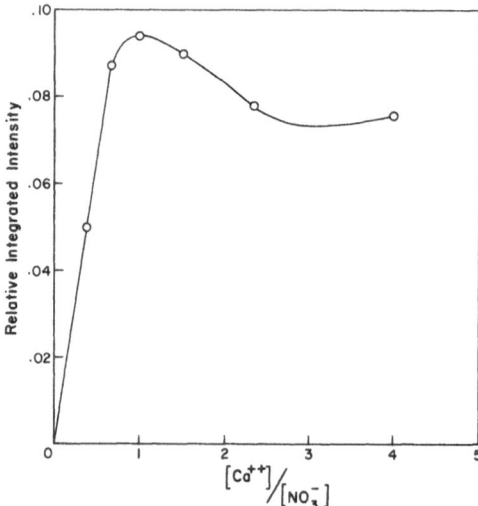

Fig. 1. Relative integrated intensity of the 740 cm$^{-1}$ band *versus* the ratio $[Ca^{+2}]/[NO_3^-]$ for aqueous solutions in which the sum $[NO_3^-] + [Ca^{+2}]$ is 5 moles/liter.

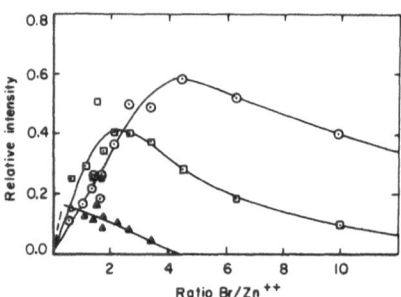

Fig. 2. Raman intensities (relative to molar $ClO_4^-$) of the three polarized bands from aqueous zinc bromide solutions in which the sum $[Zn^{+2}] + [Br^-]$ is 6 moles/liter. $\odot$, $\square$, and $\triangle$ denote the 172, 186, and 205 cm$^{-1}$ bands, respectively. Reprinted by courtesy of W. Yellin and R. A. Plane[22] from the *Journal of the American Chemical Society* by permission of the copyright owners, the American Chemical Society.

band. The $720 \, \text{cm}^{-1}$ band is still largely characteristic of free nitrate ion in the solution. The intensity of this band exhibits a pronounced minimum at the point where the concentration of perturbed nitrate is at a maximum.

If several complex ions are present in a solution, the method of continuous variations is less reliable[20] and more complicated.[18,21] If solution behavior is ideal, the intensity of a Raman line characteristic of only species $MA_n$ will be maximum at a ratio of $C_A/C_M$ given by

$$C_A/C_M = n + \frac{\sum\limits_{j=0}^{i} j(n-j)C_j}{\sum\limits_{j=0}^{i} C_j} \qquad (2)$$

where $i$ is the number of complexes of the type $MA_j$ and $C_j$ is the concentration of $MA_j$. The presence of complexes other than $MA_n$ results in a correction term [the second term on the right-hand side of equation (2)] which modifies the value of $n$ as inferred from the ratio $C_A/C_M$. The term $C_A/C_M$ will be larger than $n$ for high $n$ and lower than $n$ for low $n$. Yellin and Plane [22] have used intensity data to assign the three prominent polarized Raman bands from zinc bromide solutions (at 172, 186, and $205 \, \text{cm}^{-1}$) to the species $ZnBr_4^{-2}$, $ZnBr_2$, and $ZnBr^+$, respectively. Their Job plot is illustrated in Fig. 2.

The Job method may not always lead to reliable values of $n$, but it can serve another useful function in Raman studies. Band intensities originating from the same species should exhibit similar behavior when the composition of the solutions is varied. Thus, when the spectrum contains bands from several different species, the bands may be sorted and classified by this experimental approach.

## EQUILIBRIUM CONSTANTS

For the general expression for the formation of a complex $M_m A_n$

$$mM + nA \rightleftharpoons M_m A_n$$

the equilibrium constant $K$ may be written as follows:

$$K = \frac{[M_m A_n]}{[M]^m [A]^n} \cdot \frac{f_{M_m A_n}}{f_M^m \cdot f_A^n} = Q_c \cdot Q_f$$

where $Q_c$ is the concentration quotient written in accordance with the law of mass action and $Q_f$ is the quotient of activity coefficients. Raman spectroscopy has proven to be a powerful tool for the measurement of $Q_c$. This measurement permits calculations of $Q_f$ for those equilibria for which $K$ is known from other physical measurements (e.g., electromotive force of cells). Knowledge of $Q_f$ for many electrolyte systems of moderate to high concentration may lead to a better understanding of the concentrated electrolytic solution. Knowledge of the dependence of $Q_c$ on concentration may permit estimation of $K$ in favorable cases.

To evaluate $Q_c$, concentrations of Raman-active species are obtained from measurements of the integrated intensities of characteristic Raman bands. The integrated intensity of the band is the area between the trace as recorded by an instrument equipped with photoelectric detection* and the line drawn as a smooth continuation of the base line on either side of the band.

It is standard practice to compare the intensity of a band from the species under investigation with the intensity of a band from a reference substance of known, fixed concentration. Intensity ratios, designated "relative Raman band intensities" are then reported. This practice permits comparison of band intensities from different solutions obtained at different times in the laboratory even when the illumination from the exciting lamps has not been constant because of lamp fluctuations, lamp darkening, and electronic drift. If an internal standard is used, a constant amount of a substance which does not affect the system being studied is dissolved in each solution. Perchlorate ion has been used frequently for studies of aqueous solutions, and, except for ionic strength effects, probably does not interfere with most systems unless the perchlorate bands overlap bands of the species of interest.[24] When an external standard is employed, the spectra are obtained in a sequence such as reference–sample–reference. The relative integrated intensity of the band from the sample is then obtained by dividing the band area of the sample by the average band area of the bracketing reference. Several such areas are averaged. Care must be taken to position the cells which contain sample and reference in the same manner in the light bath for each run.

If the intensities and concentrations are to be related, the following corrections must be made to the intensity ratios:

---

* Line intensity cannot be measured with sufficient accuracy for estimation of species concentrations from the blackening of a photographic plate. For a discussion of this point see the work of Young et al.[23]

1. If more than one sample tube is used, the cells must be calibrated. The intensity ratio obtained when the same solution is examined in each of two cells provides the cell correction factor.

2. Differences in optical attenuation in the various samples will cause the data to scatter. Samples should be routinely filtered through a fine porosity filter to remove dust. If the color of the solutions (the absorbance) is changing over the region of the electromagnetic spectrum being employed in the Raman studies, serious systematic errors may occur. Empirical correction factors can be established to correct for a small, changing absorbance. Minute quantities of an absorbing species, e.g., a dye, are added to a set of solutions containing a fixed concentration of a Raman-active species. For each solution the Raman band intensity and the visible absorbance at a fixed wavelength (e.g., 4358 Å, the wavelength of the mercury exciting line) are measured. It is assumed that the Raman band intensity should be the same in all samples. The variation in intensity is related to the increased absorbance of the solution. A correction factor can then be plotted as a function of the measured absorbance. This procedure may be employed when the maximum absorbance encountered is small.

3. The correction for differences in the refractive index of the various samples is less well understood and disagreement exists as to its importance. When the refractive index of the sample changes, the amount of light entering or leaving the sample cell changes because of the change in intensity distribution between the reflected and re-fracted parts of the beam. It is difficult to obtain an empirical correction factor because substances which will markedly affect the refractive index will also interfere with the Raman-active species. One of the advantages of the use of an internal standard is the minimization of the error due to the variations of the refractive index. If the Raman sample cells are large (e.g., 50 ml volume), it is probable that the correction is never larger than a few percent. The correction is very large for microcells. (For a more detailed account, the reader is referred to other work.[74–76])

The ratio of the relative integrated intensity of a line (as obtained by the above procedure) to the molar concentration of the species generating the line is defined as the molar intensity. We denote this quantity by the symbol $J_i$, where $i$ is the line frequency; this symbol is preferred to $k$ because $k$ is often used for concentration quotients or equilibrium constants. For many species the molar intensity is essentially independent of concentration. However, nonlinearities are known and, in some cases, are a consequence of ion–ion or ion–solvent interactions.[25] It should be noted that there is a significant decrease in precision if peak heights are utilized instead of band areas. Several methods of establishing the value of the molar intensity of a band will be described below.

We will illustrate the general method of obtaining species concentrations by specific reference to several published studies. First let us consider the dissociation of the bisulfate ion in sulfuric acid solutions.

$$HSO_4^- \rightleftharpoons SO_4^{-2} + H^+$$

The sulfate ion has a strong polarized Raman band at $980\ cm^{-1}$. The molar intensity of this band can be determined from measurements of the band intensities of ammonium sulfate solutions of known composition. The linearity is nearly perfect up to about 6 moles/liter.[23] The bisulfate ion has a strong line at $1040\ cm^{-1}$. In dilute solutions of sulfuric acid, the only species present are $SO_4^{-2}$, $HSO_4^-$, hydronium ions, and water. Thus, the bisulfate ion concentration can be calculated as the difference between the stoichiometric concentration and the sulfate ion concentration. The molar intensity of the $1040\ cm^{-1}$ line can then be obtained. Again it has been found that the line intensity is directly proportional to species concentration.[23]

With the molar intensities established, the species concentrations can be computed directly from the line intensities for solutions ranging in composition from very dilute sulfuric acid to very concentrated oleum. Since the dissociation constant $K$ is well known for this system, the activity coefficient quotient has been computed (see Fig. 10 in the work of Young et al.[23]). It is not our purpose to discuss the interpretation of this interesting system. The reader may refer to the original papers.

The dissociation constant $K_2$ has recently been evaluated[26] by extrapolation of the concentration quotients obtained from Raman intensities:[23]

$$\log K_2 = \log Q_c + \log Q_f$$

The limiting law for the variation of the activity coefficients is given by the equation

$$\log Q_f = -\frac{S\sqrt{I}}{1 + A\sqrt{I}}$$

where $S$ is the limiting slope and is unambiguously defined in terms of fundamental constants (values of $S$ have been tabulated[27]), $I$ is the ionic strength of the system, and $A$ is an adjustable parameter related to the average distance of closest approach of ions in the system. From successive trials a value of $A$ equal to 0.3 was found such that $(\log Q_c + \log Q_f)$ was constant for $0 < I < 1$. The dissociation constant obtained from this extrapolation ($K_2 = 0.0102$ mole/kg) is in good agreement with values obtained from conductance, optical extinction, electromotive force of cells, and solubility studies,[26] which substantiates the reliability of the Raman method despite the difficulties of measuring Raman line intensities from low concentrations. The knowledge of the limiting slope greatly facilitates the extrapolation. Even so, care must be exercised because, as has recently been pointed out, equilibrium quotients found at high ionic strengths and thermodynamic constants obtained by extrapolation may not always refer to the same species.[28]

The molar intensity of a band characteristic of a metal complex can be obtained only if the concentration of the complex can be ascertained. When the metal concentration is maintained constant and the ligand concentration is increased, virtually all of the cation may be converted to the complex ion. Then the intensity of a characteristic line will approach a limit asymptotically with increase of the ligand-to-metal ratio. The maximum intensity should be directly proportional to the concentration of the complex ion given by the metal concentration. This requirement can be tested by obtaining the maximum intensities for several series of solutions, each series having a different constant metal-ion concentration. The molar intensity is then obtained from the slope of a plot of the maximum intensities *versus* the concentration of the complex ion (i.e., the metal ion).

This procedure is illustrated by Figs. 3 and 4. Figure 3 shows the behavior of the intensity of the 278 cm$^{-1}$ Raman band characteristic of the tetrachlorozincate (II) ion as the chloride-to-zinc ratio is increased in solutions of constant zinc concentration. The maximum intensity for each series is plotted *versus* the concentration of zinc in Fig. 4. The slope is the molar intensity of the 278 cm$^{-1}$ Raman band.[29] A similar approach has been applied to the gallium bromide system[19] and to the zinc bromide and cadmium bromide systems.[22]

In some cases, molar intensity values may be estimated by the application of Wolkenstein's theory of Raman intensities.[30,31] According to this theory, the change of mean molecular polarizability

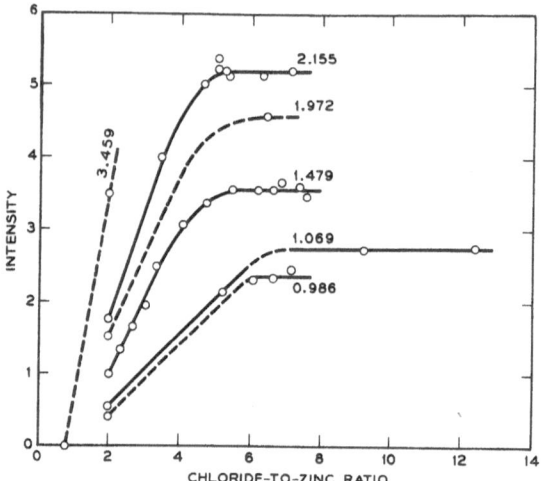

Fig. 3. Raman intensities of the 278 cm$^{-1}$ line from aqueous zinc chloride solutions as a function of the chloride-to-zinc ratio at 25°C. The curves are labeled with the zinc concentrations. (Reprinted by courtesy of the American Institute of Physics from the *Journal of Chemical Physics*.[29])

during vibration $\partial\alpha/\partial q$, which is responsible for Raman intensity, is a bond property. As an approximation then, the molar intensity is proportional to the number of M–A bonds in the complex.[22] Corrections for differences of frequency and degree of depolarization of the Raman bands can be made. Yellin and Plane[22] applied this principle and, with knowledge of the molar intensity of $ZnBr_4^{-2}$, obtained molar intensities for $ZnBr_2$ and $ZnBr^+$.

The concentration quotient $Q_c$ may also be estimated from the data obtained from the Job analysis and equation (1) given above if the molar intensity $J$ is known. Thus, replacement of $[MA_n]$ in equation (1) by $I/J$ gives

$$Q_c = \frac{I}{J}[(1-f)C - I/J]^{-1}[fC - nI/J]^{-n}$$

If the values of $I_{max}$, $f$, $C$, and $n$ from Fig. 1 and an approximate value of $J$ are substituted, a value of 0.09 (moles-liter)$^{-1}$ is obtained for $Q_c$ for the nitratocalcium species in the medium 2.5M $CaCl_2$–2.5M $NH_4NO_3$.[14]

Useful information concerning equilibria may sometimes be obtained from equations involving molar intensities and species concentrations. An integrated intensity may include contributions from more than one species because of fortuitous band overlap. If Beer's law is assumed,

$$I = \sum_i J_i C_i$$

where $i$ is the number of species contributing to the intensity at the measured frequency. If, for the same species, other measurable bands exist in the spectrum, ratios of measured intensities may lead to

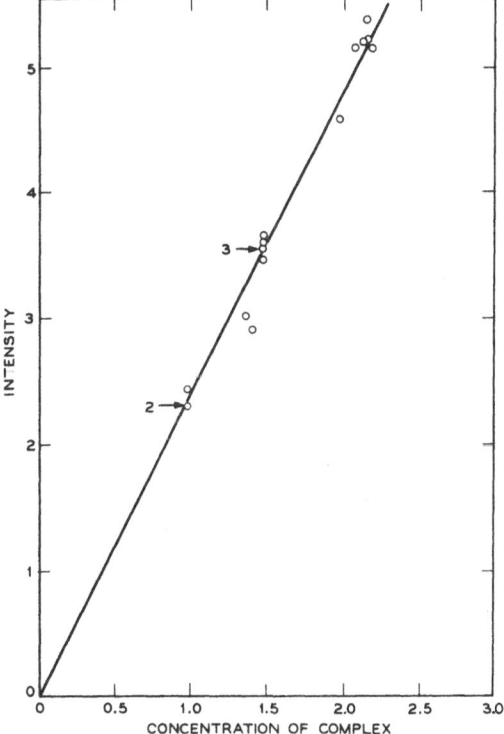

Fig. 4. Relation between the intensity of the 278 cm$^{-1}$ line and the concentration (moles/liter) of the complex ion. The numbers 2 and 3 indicate the presence of two and three points, respectively, at the positions marked. (Reprinted by courtesy of the American Institute of Physics from the *Journal of Chemical Physics*.[29])

equations the analysis of which permits distinctions to be made between alternative models. Thus, Walrafen[32] devised an intensity analysis which required the plot of two intensity ratios to be a straight line if one band had an intensity contribution from two different species. The observed linearity confirmed the band assignment and yielded molar intensity ratios.

## THERMODYNAMIC FUNCTIONS

Enthalpy and entropy changes for association reactions can be obtained from knowledge of the temperature dependence of the equilibrium constants. Control over a wide range of temperatures is easily accomplished for Raman studies. However, Raman spectroscopy provides only concentration quotients for relatively concentrated solutions, and extrapolations such as the one described above for the bisulfate equilibrium may not be possible. In general, activity coefficients are not known and extrapolations involving ionic strength cannot be considered reliable and may not be practical. This limits the value of the thermodynamic data, although the values may be adequate to reveal certain trends for series of complexes. Nixon and Plane[19] have estimated values of $\Delta H$ and $\Delta S$ for the association of $GaBr_4^-$.

## KINETICS

The essentially constant illumination over long periods of time provided by lamps now commercially available makes practical the study of Raman band intensities as a function of time. Martin[33] has employed this technique to measure the rate of exchange of the hydrogen atom bonded to phosphorus in phosphorous acid. Upon addition of $H_3PO_3$ to $D_2O$, lines at 750 and 1793 $cm^{-1}$ appear and increase in intensity, while lines at 1024 and 2457 $cm^{-1}$ decrease in intensity. The 750 and 1793 $cm^{-1}$ bands are attributed to the P–D deformation and stretch, respectively. The 1024 and 2457 $cm^{-1}$ bands are the analogs for the P–H group. The growth of the 750 and 1793 $cm^{-1}$ bands was followed relative to the intense 940 $cm^{-1}$ band, which being common to both $H_3PO_3$ and $D_3PO_3$ was used as an internal reference. The half-life for the exchange was found to be about 200 min at 23°C.

Kreevoy and Mead[34] have explored the usefulness of measuring line broadening to measure fast reaction rates (half-times from $\sim 10^{-11}$

to $\sim 10^{-14}$ sec). The broadening of the 1433 cm$^{-1}$ band was used to determine the first-order rate constant for proton transfer from trifluoroacetic acid to the trifluoroacetate ion. These two studies point up the potential of a still unpursued area of research.

# PART 2

## RAMAN SPECTRA OF SIMPLE COMPLEX IONS

In this section the Raman spectra of some classes of simple complex ions are presented and discussed. No attempt has been made to give an exhaustive review of the literature or to give complete references. The references cited should, however, lead the reader to most of the literature concerning the species of interest. For greater coverage the reader is referred to a recent text by Nakamoto.[35]

### Aquo Complexes

An exciting development in the study of aqueous solutions of electrolytes has been the observation and assignment of Raman bands to ion–water vibrational modes. Mathieu[36] reported a broad polarized band in the region 371–388 cm$^{-1}$ in the Raman spectra of six crystalline zinc salts known to contain $Zn(H_2O)_6^{+2}$. He observed a line at 387 cm$^{-1}$ for aqueous 7N $Zn(NO_3)_2$ and a line at 370 cm$^{-1}$ for aqueous 5N $ZnSO_4$. The band was ascribed to the hexaquozinc(II) ion. Mathieu also reported that solutions of magnesium nitrate show a line at 380 cm$^{-1}$ and that $[Mg(H_2O)_6]SO_4 \cdot H_2O$ has a broad polarized band at 382 cm$^{-1}$. Similarly, solutions and hydrated crystals of copper nitrate have a band in the region 390–435 cm$^{-1}$. Lafont[37] extended the Raman studies of crystalline hydrates.

The early work was extended by da Silveira, Marques, and Marques[38] who studied the bands generated by magnesium, zinc, aluminum, and beryllium salts dissolved in both light and heavy water. They ascribed bands at 363, 315, and 240 cm$^{-1}$ to $[Mg(H_2O)_6]^{+2}$; 524, 447, and 340 cm$^{-1}$ to $[Al(H_2O)_6]^{+3}$; 389, 335, and 230 cm$^{-1}$ to $[Zn(H_2O)_6]^{+2}$; and 350 and 536 cm$^{-1}$ to $[Be(H_2O)_4]^{+2}$.

Irish, McCarroll, and Young[39] reported the broad, weak polarized 390 cm$^{-1}$ band in aqueous zinc chloride solutions. This band disappeared on addition of an excess of chloride ion because all the hexaquozinc(II) ion is virtually converted to the chlorozincate complex. Solutions of zinc nitrate generate the bands 394, 226, and 160–

## TABLE II
### Raman Spectra of Some Halo Complex Ions

| Species | Point group | Fundamental frequencies (cm$^{-1}$) | | | | | References |
|---|---|---|---|---|---|---|---|
| | | $\nu_1$ | $\nu_2$ | $\nu_3$ | $\nu_4$ | $\nu_5$ | |
| **Group IB** | | | | | | | |
| $AuCl_4^-$ | $D_{4h}$ | 347 | 171 | | 324 | | 41, 42 |
| $AuBr_4^-$ | $D_{4h}$ | 212 | 102 | | 196 | | 41 |
| **Group IIB** | | | | | | | |
| $ZnCl^+$ | $C_{\infty v}$ | >305 | | | | | 29 |
| $ZnCl_2$ | $D_{\infty h}$ | 305 | | | | | 45, 47, 29 |
| $ZnCl_4^{-2}$ | $D_{4h}$ | 278 | 110 | | | | 45, 29 |
| $ZnBr^+$ | $C_{\infty v}$ | 205 | | | | | 22 |
| $ZnBr_2$ | $C_{2v}$ | 186 | | | | | 22 |
| $ZnBr_4^{-2}$ | $T_d$ | 172 | 61 | 210 | 82 | | 22 |
| $ZnI^+$ | $C_{\infty v}$ | 163 | | | | | 43 |
| $ZnI_2$ | | 140 | | | | | 43 |
| $ZnI_4^{-2}$ | $T_d$ | 122 | 44 | 170 | 62 | | 43 |
| $CdBr_4^{-2}$ | $T_d$ | 166 | 53 | 183 | 62 | | 22, 44 |
| $CdI_4^{-2}$ | $T_d$ | 117 | 36 | 145 | 44 | | 45 |
| $HgI_4^{-2}$ | $T_d$ | 126 | 35 | | 41 | | 46 |
| **Group IIIA** | | | | | | | |
| $AlCl_4^-$ | $T_d$ | 349 | 146 | 575 | 180 | | 48 |
| $GaCl_4^-$ | $T_d$ | 346 | 114 | 386 | 149 | | 49 |
| $GaBr_4^-$ | $T_d$ | 210 | 71 | 278 | 102 | | 50 |
| $GaI_4^-$ | $T_d$ | 145 | 52 | 222 | 73 | | 51 |
| $InCl_4^-$ | $T_d$ | 321 | 89 | 337 | 112 | | 52 |
| $InBr_4^-$ | $T_d$ | 197 | 55 | 239 | 79 | | 53 |
| $InI_4^-$ | $T_d$ | 139 | 42 | 185 | 58 | | 51 |
| $TlCl^{+2}$ | $C_{\infty v}$ | 327 | | | | | 73 |
| $TlCl_2^+$ | $D_{\infty h}$ | 320 | | | | | 73 |
| $TlCl_3$ | $D_{3h}$? | 313 | | | | | 73 |
| $TlCl_4^-$ | $T_d$? | 305 | 81 | | | | 73 |
| $TlCl_6^{-3}$? | $O_h$ | 273 | | | | | 73 |
| $TlBr_4^-$ | $T_d$ | 190 | 51 | 209 | 64 | | 54 |
| **Group IVA** | | | | | | | |
| $GeF_6^{-2}$ | $O_h$ | 627 | 454 | | | 318 | 11 |
| $SnCl_6^{-2}$ | $O_h$ | 311 | 229 | | | 158 | 55 |
| $PbCl_6^{-2}$ | $O_h$ | 285 | 215 | | | 137 | 56 |
| $SnBr_6^{-2}$ | $O_h$ | 185 | 138 | | | 95 | 55 |

176 cm$^{-1}$. The 394 cm$^{-1}$ band was also observed in the spectrum of a 3M solution of zinc perchlorate.

Goggin and Woodward[39] ascribed a band at 463 cm$^{-1}$ from aqueous solutions of methyl mercuric perchlorate to the mercury–water stretch of the species $[CH_3Hg-OH_2]^+$.

Hester and Plane[40] reported a polarized band in the region 360–400 cm$^{-1}$ from solutions of copper, zinc, mercury, magnesium, indium, gallium (475), and thallium (470). Aquo $Tl^{+3}$ has been studied by Spiro.[73]

The broad, diffuse nature of the polarized band has not been adequately explained, but it may reflect the rapid exchange of water molecules between the coordination sphere of the ion and the bulk solvent.

## Halo Complexes

Table II lists the fundamental frequencies for some halo complexes of families IB, IIB, IIIA, and IVA. The tetrahedral complexes exhibit a characteristic four-line spectrum portrayed in Fig. 5 for $ZnI_4^{-2}$ present in a $ZnI_2$–LiI solution. The polarization of the intense 122 cm$^{-1}$ band is evident from comparison of parts (b) and (c), obtained by using exciting light polarized parallel and perpendicular to the axis of the exciting tube, respectively. The contours of Fig. 5 were obtained from a photoelectric recording by transferring the difference in amplitude between the curved base line and the trace to a horizontal base line. In the absence of excess lithium iodide, three polarized bands are present in the spectrum of an aqueous zinc iodide solution, $viz.$, 122, 140, and 163 cm$^{-1}$. Three polarized bands also occur for $ZnBr_2$ solutions—172, 186, and 205 cm$^{-1}$. The spacing between the lines decreases as the mass of the ligand decreases, and for $ZnCl_2$ solutions, a single, broad asymmetric envelope at 284 cm$^{-1}$ is observed. A difference of opinion exists on the assignment of the three polarized vibrations,[57,58] although it is clear that they characterize species with ligand-to-metal ratios of four and lower. The assignment of the vibrations to $ZnX_2$ and $ZnX^+$ is in accord with the Job analysis of Yellin and Plane[22] discussed above. Similar band overlap probably occurs in other systems, e.g., the haloindium complexes[52] and the chlorocomplexes of thallium (III).[73]

Comparison of the spectra of $ZnI_4^{-2}$ and $ZnBr_4^{-2}$ with that of $ZnCl_4^{-2}$ is also of interest. For the latter, only two lines are observed.[29]

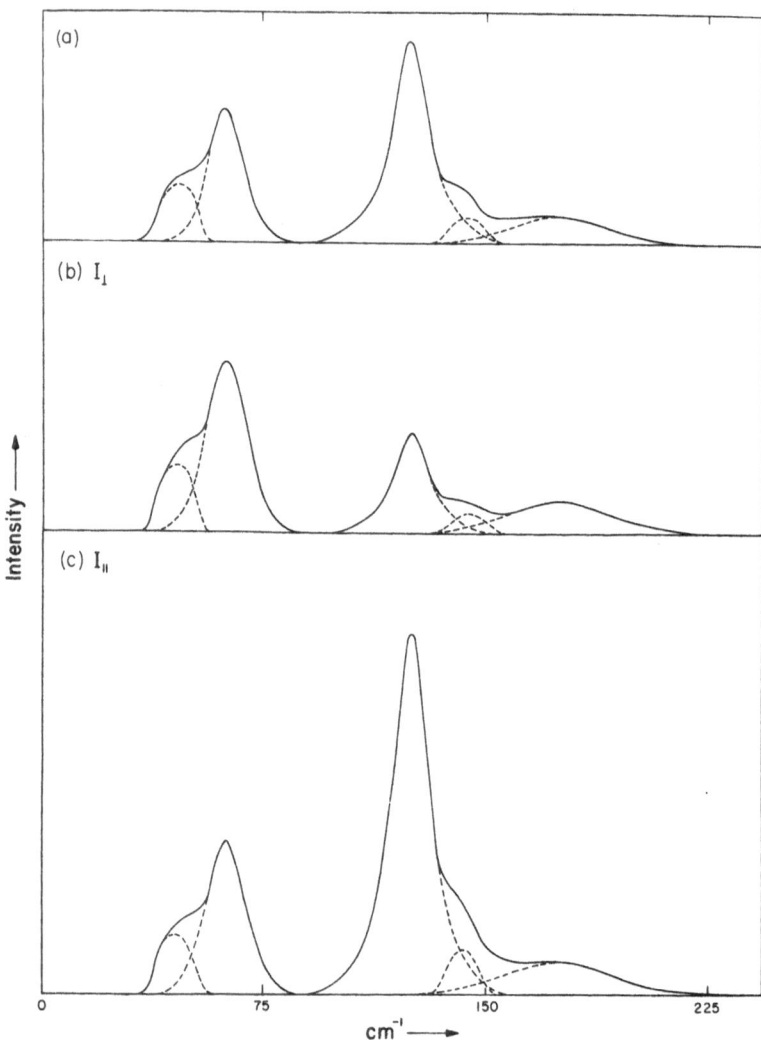

Fig. 5. The contours of the spectra of an aqueous solution of $ZnI_2$ and LiI after the ordinates between base line and trace have been transferred to a horizontal base line. (a) Unpolarized exciting light. (b) Exciting light polarized parallel to the axis of the sample tube. (c) Exciting light polarized perpendicular to the axis of the sample tube. (Published by courtesy of T. F. Young and P. M. Vollmar.)

However, if the symmetry of $ZnCl_4^{-2}$ is $T_d$, four lines should be observed. Furthermore, one would expect a greater separation of the $v_2$ and $v_4$ bands for the chloride than for the bromide or iodide (see Fig. 6). For the series illustrated in Fig. 6, for a given central atom, as the mass of the ligand decreases, the separation $(v_4 - v_2)$ increases; for a given ligand, as the mass of the central atom in the same group of the periodic table decreases, the separation increases. Because the two frequencies of $ZnBr_4^{-2}$ and $ZnI_4^{-2}$ are resolved and observable, even though closer to the exciting line than those of the corresponding chloride, one would expect the two vibrations of $ZnCl_4^{-2}$ to be readily observed. The fact that only one vibration is observed strongly suggests that the species does not have tetrahedral symmetry. Octahedral symmetry with two water molecules occupying the two axial positions has been suggested for $ZnCl_4^{-2}$.[29] This geometry is consistent

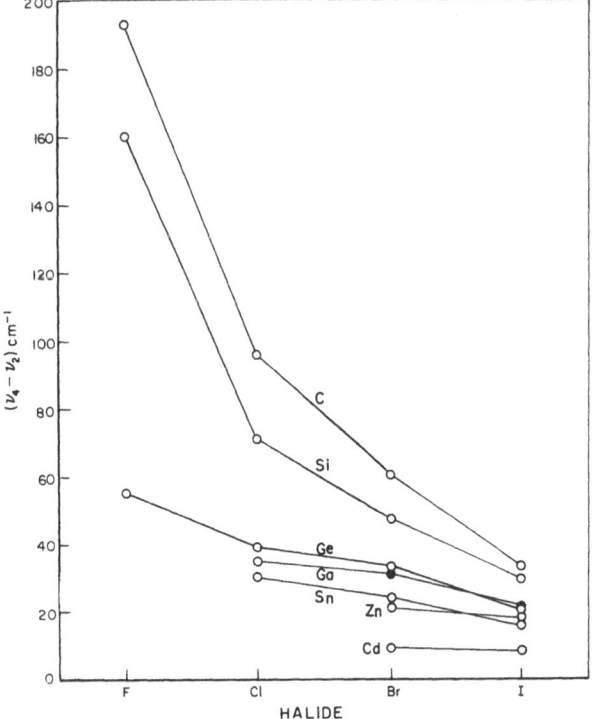

Fig. 6. The difference $(v_4 - v_2)$ for some tetrahalo species with $T_d$ symmetry.

with the size of the chloride ion, but is not probable for complex ions involving the larger bromide and iodide ions. Infrared spectra would be useful for the clarification of this structure.

Woodward and Roberts[59] noted that for a regular tetrahedral molecule $MA_4$ the ratio $v_4/v_2$ passes from being greater than one to being less than one as the mass ratio $m_M/m_A$ increases. For most $MA_4$ molecules, $v_4/v_2 > 1$. For $OsO_4$, $m_M/m_A$ is $\sim 12$, and, from consideration of the simple valence force field of Heath and Linnett,[60] the approximate equality of $v_4$ and $v_2$ was predicted to account for the failure to observe two resolved bands. Busey and Keller[61] extended these calculations to $MoO_4^{-2}$ (aq) and $CrO_4^{-2}$ (aq) and noted that the separations should be observable. They concluded that the equations were not completely satisfactory for these ions and suggested that the intensity of the $v_2(E)$ bending vibration is sufficiently weak and the line sufficiently broad so that it is not detected in aqueous solution. By analogy with $GaCl_4^-$, for which four Raman lines are observed, the mass-ratio argument lends some support to the postulate that the symmetry of $ZnCl_4^{-2}$ is not $T_d$. However, low intensity of the deformation band cannot be precluded at this time (cf., Spiro[73]).

## Sulfates and Perchlorates

Raman spectra of both the sulfate and the perchlorate ions are consistent with $T_d$ symmetry. Hester and Plane[24] have reported the Raman spectra of a wide range of metal sulfates and perchlorates, obtained as part of a study of metal–oxyanion associations in aqueous solution. No spectral evidence was obtained for association, with the exception of indium sulfate.[62]

## Nitrates

A free nitrate ion of point group $D_{3h}$ would be expected to have the following vibration spectrum: $v_1(A_1')$ 1048 cm$^{-1}$ $(R)$; $v_2(A_2'')$ 825 cm$^{-1}$ $(I)$; $v_3(E')$ 1384 cm$^{-1}$ $(R, I)$; $v_4(E')$ 719 cm$^{-1}$ $(R, I)$. This spectrum is rarely if ever observed for an aqueous solution of a nitrate salt.[63] The molar intensity, frequency, and band width of the 1048 cm$^{-1}$ band depend on concentration and cation.[25] This band (sometimes shifted to lower frequency) is infrared-active, even for salts of $NH_4^+$, $K^+$, and $Na^+$.[14] The intensity of this band in the infrared is often greater than that of the infrared-active 825 cm$^{-1}$ band. The 1384 cm$^{-1}$ band degeneracy is removed, even for solutions of $KNO_3$, and two bands at $\sim 1350$ and $1450\,\text{cm}^{-1}$ are observed. This split has been attributed to ionic interaction.[64] The degeneracy of the $719\,\text{cm}^{-1}$ band is sometimes removed. Hester and Plane[13] ascribed a band at $740\,\text{cm}^{-1}$ from aqueous solutions of calcium nitrate to the $\text{CaNO}_3^+$ ion, a stoichiometry substantiated by Fig. 1. Concurrently for those solutions, a band at $717\,\text{cm}^{-1}$ was observed and has been assigned to the unbound nitrate ion.[13]

For aqueous solutions of copper nitrate and thorium nitrate, two polarized bands have been observed in the $1050\,\text{cm}^{-1}$ region ($1048$ and $1024\,\text{cm}^{-1}$ for copper and $1050$ and $1034\,\text{cm}^{-1}$ for thorium).[63] Hester and Plane[24] have summarized data for a number of metal nitrates and attributed new frequencies to nitrate complexes with $C_{2v}$ point-group assignment, indicative of complexing through oxygen atoms of the $\text{NO}_3^-$ ions. They suggest that the magnitude of the splitting of the $\nu_3(E')$ line of the $D_{3h}$ point group into $\nu_4(B_1)$ and $\nu_1(A_1)$ lines of the $C_{2v}$ point group is a measure of the dissymmetry in the nitrate group. This criterion predicts the following order of nitrate deformations:

$$Th^{+4} > In^{+3} > Cu^{+2} > Hg^{+2} > Ce^{+3} > Ca^{+2} > Zn^{+2}, Al^{+3},$$

$$Ag^+ > Na^+, K^+, NH_4^+$$

The absence of a metal–nitrate vibration may indicate that the ionic interaction is through intervening water molecules.[14] Some disagreement apparently exists on the number of observed bands in the region $1350\text{–}1450\,\text{cm}^{-1}$.

The spectra of solutions of methyl mercuric nitrate in benzene suggest that only the undissociated molecules $CH_3Hg{-}ONO_2$ are present, whereas in water there is an equilibrium between these molecules and the ions $[CH_3Hg{-}OH_2]^+$ and $NO_3^-$.[39] A polarized frequency at $292\,\text{cm}^{-1}$ is assigned to the Hg–O stretch of the $CH_3Hg{-}ONO_2$ molecule. Except for the absence of a Raman vibration in this low-frequency region, there are many similarities between the spectra of $Cu(NO_3)_2$ and $CH_3HgNO_3$. An analogous complex has been formed by mercury(II), acetone, and nitrate ion in aqueous solution.[65] The intensity of the spectrum of the nitrate group bound to either acetone–Hg or methyl–Hg is significantly greater than that when bound to the hydrated mercuric ion.

## TABLE III
### Raman Spectra of Some Oxyanions and Related Species

| Species | Point group | | | Raman bands (cm$^{-1}$) | | | | | Reference |
|---|---|---|---|---|---|---|---|---|---|
| $OsO_4$ | $T_d$ | | 335 | — | 965(p)* | 954 | | | 59 |
| $ReO_4^-$ | $T_d$ | | 332 | — | 971(p) | 916 | | | 66 |
| $WO_4^{-2}$ | $T_d$ | | 324 | — | 931(p) | 833 | | | 67 |
| $MoO_4^{-2}$ | $T_d$ | | 318 | — | 897(p) | 841 | | | 61 |
| $TcO_4^-$ | $T_d$ | | 325 | — | 912(p) | — | | | 61 |
| $VO_4^{-3}$ | $T_d$ | | 345 | 480 | 870(p) | 825 | | | 67 |
| $CrO_4^{-2}$ | $T_d$ | | 348 | 368 | 847(p) | 884 | | | 68 |
| $AsO_4^{-3}$ | $T_d$ | | 349 | 463 | 837(p) | 878 | | | 67 |
| $IO_4^-$ | $T_d$ | | 256 | 325 | 791(p) | 853 | | | 69 |
| $SO_4^-$ | $T_d$ | | 451 | 613 | 981(p) | 1104 | | | 70 |
| $HSO_4^-$ | $C_s$† | | 417 | 593 | 885(p) | 1050(p) | 1230 | 1340 | 70 |
| $H_2SO_4$ | $C_{2v}$† | 390 | 420 | 563 | 913(p) | 972 | 1135(p) | 1365 | 70 |
| $SeO_4^{-2}$ | $T_d$ | | 348 | 414 | 837(p) | 875 | | | 70 |
| $HSeO_4^-$ | $C_s$† | | 320 | 395 | 740(p) | 862(p) | 915 | 945 | 70 |
| $H_2SeO_4$ | $C_{2v}$† | 290–305 | 362 | ~400 | 754(p) | 785 | 898(p) | 975 | 70 |
| $ClO_4^-$ | $T_d$ | | 460 | 630 | 935(p) | 1050–1070 | | | 71 |
| $Al(OH)_4^-$ | $T_d$ | | 310 | — | 615(p) | — | | | 72 |
| $Zn(OH)_4^-$ | $T_d$ | | 300 | — | 470(p) | — | | | 72 |

\* The letter $p$ denotes a polarized line.
† Considered "five-atomic."

## Some Oxyanions and Related Species

Raman vibrations of some oxyanions and related species are presented in Table III. For the tetrahedral species the vibrations are listed in the order: $v_2(E)$, $v_4(F_2)$, $v_1(A_1)$, $v_3(F_2)$. The characteristic four-line spectrum for a species with tetrahedral symmetry is not observed for the first five entries. Either certain vibrations are accidentally degenerate[59] or the bands are too weak to be observed.[61] The first six entries are also irregular in that $v_3$ is less than $v_1$.

## REFERENCES

1. G. Herzberg, *Molecular Spectra and Molecular Structure, Vol. II, Infrared and Raman Spectra of Polyatomic Molecules*, D. Van Nostrand Company, Inc., Princeton, New Jersey, 1945.
2. F. A. Cotton, *Chemical Applications of Group Theory*, Interscience Publishers, New York, 1963, Chapter 3.
3. *Ibid.*, Chapter 9.
4. J. E. Rosenthal and G. M. Murphy, *Rev. Mod. Phys.* **8**: 317 (1936).
5. A. G. Meister, F. F. Cleveland, and M. J. Murray, *Am. J. Phys.* **11**: 239 (1943).
6. E. B. Wilson, Jr., *J. Chem. Phys.* **2**: 432 (1934).

7. D. M. Sweeny, I. Nakagawa, S. Mizushima, and J. V. Quagliano, *J. Am. Chem. Soc.* **78**: 889 (1956); C.W.F.T. Pistorius, *Z. Physik. Chem. (Frankfurt)* **23**: 206 (1960).
8. R. S. Halford, *J. Chem. Phys.* **14**: 8 (1946).
9. R. S. Halford and O. A. Schaeffer, *J. Chem. Phys.* **14**: 141 (1946).
10. H. Winston and R. S. Halford, *J. Chem. Phys.* **17**: 607 (1949).
11. J. E. Griffiths and D. E. Irish, *Inorg. Chem.* **3**: 1134 (1964).
12. J. M. Smithson and R. J. P. Williams, *J. Chem. Soc.* 457 (1958).
13. R. E. Hester and R. A. Plane, *J. Chem. Phys.* **40**: 411 (1964).
14. D. E. Irish and G. E. Walrafen, *J. Chem. Phys.* **46** (January 1967), in press.
15. J. H. B. George, J. A. Rolfe, and L. A. Woodward, *Trans. Faraday Soc.* **49**: 375 (1953).
16. C. B. Baddiel, M. J. Tait, and G. J. Janz, *J. Phys. Chem.* **69**: 3634 (1965).
17. P. Job, *Ann. Chim. (Paris)* **9**: 113 (1928).
18. W. C. Vosburgh and G. R. Cooper, *J. Am. Chem. Soc.* **63**; 437 (1941).
19. J. Nixon and R. A. Plane, *J. Am. Chem. Soc.* **84**: 4445 (1962).
20. F. J. C. Rossotti and H. Rossotti, *The Determination of Stability Constants*, McGraw-Hill Book Co., Inc., New York, 1961, p. 50.
21. L. I. Katzin and E. Gebert, *J. Am. Chem. Soc.* **72**: 5455 (1950).
22. W. Yellin and R. A. Plane, *J. Am. Chem. Soc.* **83**: 2448 (1961).
23. T. F. Young, L. F. Maranville, and H. M. Smith, "Raman Spectral Investigations of Ionic Equilibria in Solutions of Strong Electrolytes," in: W. J. Hamer (ed.), *The Structure of Electrolytic Solutions*, John Wiley and Sons, Inc., New York, 1959, p. 35.
24. R. E. Hester and R. A. Plane, *Inorg. Chem.* **3**: 769 (1964).
25. P. M. Vollmar, *J. Chem. Phys.* **39**: 2236 (1963).
26. T. F. Young and D. E. Irish, "Solutions of Electrolytes," in: H. Eyring (ed.), *Annual Review of Physical Chemistry*, Vol. 13, Annual Reviews, Inc., Palo Alto, California, 1962, p. 435.
27. H. S. Harned and B. B. Owen, *The Physical Chemistry of Electrolytic Solutions*, third edition, Reinhold Publishing Corp., New York, 1958, p. 165.
28. E. Hogfeldt, *Acta Chem. Scand.* **17**: 785 (1963).
29. D. E. Irish, B. McCarroll, and T. F. Young, *J. Chem. Phys.* **39**: 3436 (1963).
30. M. Wolkenstein, *Compt. Rend. Acad. Sci. U.R.S.S.* **32**: 185 (1941).
31. G. W. Chantry and R. A. Plane, *J. Chem. Phys.* **32**: 319 (1960).
32. G. E. Walrafen, *J. Chem. Phys.* **42**: 485 (1965).
33. R. B. Martin, *J. Am. Chem. Soc.* **81**: 1574 (1959).
34. M. M. Kreevoy and C. A. Mead, *J. Am. Chem. Soc.* **84**: 4596 (1962).
35. K. Nakamoto, *Infrared Spectra of Inorganic and Coordination Compounds*, John Wiley and Sons, Inc., New York, 1963.
36. J. P. Mathieu, *Compt. Rend.* **231**: 896 (1950).
37. R. L. Lafont, *Compt. Rend.* **244**: 1481 (1957).
38. A. da Silveira, M. A. Marques, and N. M. Marques, *Compt. Rend.* **252**: 3983 (1961); *Mol. Phys.* **9**: 271 (1965).
39. P. L. Goggin and L. A. Woodward, *Trans. Faraday Soc.* **58**: 1495 (1962).
40. R. E. Hester and R. A. Plane, *Inorg. Chem.* **3**: 768 (1964).
41. H. Stammreich and R. Forneris, *Spectrochim. Acta.* **16**: 363 (1960).
42. J. D. S. Goulden, A. Maccoll, and D. J. Millin, *J. Chem. Soc.* 1635 (1950).
43. M. Delwaulle, *Compt. Rend.* **240**: 2132 (1955).
44. J. A. Rolfe, D. E. Sheppard, and L. A. Woodward, *Trans. Faraday Soc.* **50**: 1275 (1954).
45. M. L. Delwaulle, *Bull. Soc. Chim. France* 1294 (1955).
46. D. A. Long and J. Y. H. Chau, *Trans. Faraday Soc.* **58**: 2325 (1962).
47. Z. Kecki, *Spectrochim. Acta* **18**: 1165 (1962).

48. H. Gerding and H. Hautgraaf, *Rec. Trav. Chim.* **72**: 21 (1953).
49. L. A. Woodward and A. A. Nord, *J. Chem. Soc.* 3721 (1956).
50. *Ibid.*, 2655 (1955).
51. L. A. Woodward and G. H. Singer, *J. Chem. Soc.* 716 (1958).
52. L. A. Woodward and M. J. Taylor, *J. Chem. Soc.* 4473 (1960).
53. L. A. Woodward and P. T. Bill, *J. Chem. Soc.* 1699 (1955).
54. M. L. Delwaulle, *Compt. Rend.* **238**: 2522 (1954).
55. L. A. Woodward and L. E. Anderson, *J. Chem. Soc.* 1284 (1957).
56. J. A. Creighton and L. A. Woodward, *Trans. Faraday Soc.* **58**: 1077 (1962).
57. D. F. C. Morris, E. L. Short, and D. N. Waters, *J. Inorg. Nucl. Chem.* **25**: 975 (1963).
58. D. F. C. Morris, E. L. Short, and D. N. Slater, *Electrochim. Acta* **8**: 289 (1963).
59. L. A. Woodward and H. L. Roberts, *Trans. Faraday Soc.* **52**: 615 (1956).
60. D. F. Heath and J. W. Linnett, *Trans. Faraday Soc.* **44**: 561, 878 (1948).
61. R. H. Busey and O. L. Keller, Jr., *J. Chem. Phys.* **41**: 215 (1964).
62. R. E. Hester, R. A. Plane, and G. E. Walrafen, *J. Chem. Phys.* **38**: 249 (1963).
63. J. P. Mathieu and M. Lounsbury, *Discussions Faraday Soc.* **No. 9,** 196 (1950).
64. H. Lee and K. Wilmshurst, *Australian J. Chem.* **17**: 943 (1964).
65. R. R. Miano and R. A. Plane, *Inorg. Chem.* **3**: 987 (1964).
66. H. H. Cloassen and A. J. Zielen, *J. Chem. Phys.* **22**: 707 (1954).
67. H. Siebert, *Z. Anorg. Allgem. Chem.* **275**: 225 (1954).
68. D. Bassi and O. Sala, *Spectrochim. Acta* **12**: 403 (1958).
69. H. Siebert, *Z. Anorg. Allgem. Chem.* **273**: 21 (1953).
70. G. E. Walrafen, *J. Chem. Phys.* **39**: 1479 (1963).
71. H. Colm, *J. Chem. Soc.* 4282 (1952).
72. E. R. Lippincott, J. A. Psellos, and M. C. Tobin, *J. Chem. Phys.* **20**: 536 (1952).
73. T. G. Spiro, *Inorg. Chem.* **4**: 731 (1965).
74. H. J. Bernstein and G. Allen, *J. Opt. Soc. Am.* **45**: 237 (1955).
75. D. G. Rea, *J. Opt. Soc. Am.* **49**: 90 (1959).
76. D. D. Tunnicliff and A. C. Jones, *Spectrochim. Acta* **18**: 579 (1962).

# Index

**251**